Volume One: Futuristics: Looking Ahead

Volume Two: America: Moving Ahead

Volume Three: Getting Personal: Staying Ahead

Volume Four: Moving Along: Far Ahead

Courtesy of The Venus Project
Designed by Jacque Fresco and Roxanne Meadows

Tackling Tomorrow Today

Volume Three

Getting Personal: Staying Ahead

Edited by Arthur B. Shostak, Ph.D.
Professor Emeritus of Sociology,
Department of Culture and Communications
Drexel University, Philadelphia, PA 19104

CHELSEA HOUSE PUBLISHERS

VP, New Product Development Sally Cheney
Director of Production Kim Shinners
Creative Manager Takeshi Takahashi
Manufacturing Manager Diann Grasse

Staff for TACKLING TOMORROW TODAY

Executive Editor Lee Marcott
Editor Christian Green
Production Editor Noelle Nardone
Photo Editor Sarah Bloom
Series and Cover Designer Takeshi Takahashi
Layout EJB Publishing Services

A Haights Cross Communications Company

http://www.chelseahouse.com

First Printing

9 8 7 6 5 4 3 2 1

Library of Congress Cataloging-in-Publication Data

Tackling tomorrow today / edited by Arthur B. Shostak.
p. cm.
Includes bibliographical references and index.
ISBN 0-7910-8401-9 (v. 1) -- ISBN 0-7910-8402-7 (v. 2) -- ISBN 0-7910-8403-5 (v. 3) -- ISBN 0-7910-8404-3 (v. 4) 1. Twenty-first century--Forecasts. 2. Technology and civilization. I. Shostak, Arthur B.
CB161.T33 2004
303.49'09'05--dc

2004016198

■

Dedicated to forecasters,
prominent and unsung alike,
who help us see further,
imagine more,
prepare better
and
savor life's extraordinary possibilities.

■

There is nothing permanent except change.
—Heraclitus

■

To keep our faces toward change
and behave like free spirits
in the presence of fate
is strength undefeatable.
—Helen Keller

ACKNOWLEDGMENTS

Sixteen high school students from six states and the District of Columbia volunteered to critique well over sixty candidate essays and help me chose fifty-eight for the four volumes in this series. Their cogent and insightful feedback (266 brief reviews) can be found at the rear of each volume, and it makes clear my considerable debt to them: Amelia Adams, Mike Antonelli, Erin Bauerle, Patricia Marie Borrell, Andrew Crandall, Alex Dale, Tom Dunn, Rebecca Henderson, Mara James, Sarah Konner, Ginger Lemon, Kelly Ramirez, Dalea Reichgott, Benjamin Samuels, Brittany Tracy, and Jessica Varzaly. Special thanks go to Alex, Dalea, Mike, and others for survey responses, and for sheer output alone, to Tom, Jessica, Benjamin, Mara, Patricia, Ginger, Andrew, and Alex.

Plainly, much appreciation is owed the forty-three writers of the series' fifty-eight original essays; busy people who took time to share creative ideas and earnest feelings about our choices in making probable, possible, preferable, and preventable futures.

Several contributors (Glenn, Jeff, Joe, Josh, Marilyn, Nat, Roger, Sohail, and Tom) commented usefully on the essays of others. John Smart secured remarkable artwork for his two essays from Cris Dornaus. Marvin Cetron, Nat Irvin, Mel Konner, Robert Merikangas, and Patrick Salsbury ably adapted essays. Ann Coombs provided special research material of great value. While they did not write essays, Daniel Shostak did provide insightful discussion questions, as did

Nada Khader. Jacque Fresco and Roxanne Meadows shared their extraordinary artwork.

Many whose ideas are not aired directly in the book nevertheless made a vital contribution. Stevi Baggert, Connie Cordovilla, Judith Czigler, Alexander Friedlander, Thad McKenna, Adrienne Redd, and Emily Thorne helped recruit high school volunteers. Todd R. Grube identified editorial cartoons of high quality. Peggy Dominy, an Information Services Librarian at Drexel University, found hard-to-locate missing data. And, of course, there were many others whom I trust will forgive my regrettable memory lapse.

As before in the case of five books I edited in 2003/2004 on 9/11 and the Iraq War (DEFEATING TERRORISM/DEVELOPING DREAMS), the staff of Chelsea House did an especially fine job meeting some rather complex challenges, with special thanks going to the series editor, Christian Green.

For more than a quarter of a century, my wife, Lynn Seng, has contributed ideas of great value, reviews of keen insight, and support without which I would accomplish far less. Her belief in this project, and her love and smile, make all the difference.

Finally, I would like to acknowledge YOUR unique contribution, for it is ultimately only as you—and other readers—ponder and act on the book's many ideas that this volume can help us craft a world tomorrow that increasingly honors us all.

Table of Contents

Introduction

... the future is certain to be outrageous by any standards of today. Projections that are merely reasonable are certain to be wrong.
—Douglas S. Robertson,
The New Renaissance: Computers and the Next Level of Civilization

Our goal in this third volume, *Getting Personal: Staying Ahead*, is to turn research attention on ourselves. In volume one, at an introductory level, we looked at the field of futuristics; in volume two, at some of America's major policy choices. And in the closing volume of this open-ended series (TACKLING TOMORROW TODAY), we will move on to examine some very "iffy" possibilities (nanotechnology, etc.). Here, however, we step center stage.

Three essays in Part One explore such intimate matters as relationships of the heart, costuming, and family life. The first essay asks: What might our options be tomorrow as we seek new friends, and are eager to love and be loved in return? The second essay looks into the world of fashion, and forecasts many exciting twists and turns. The third essay answers some searching questions about the families we may form tomorrow, and highlights consistencies and differences with that which is familiar nowadays.

In Part Two, we hone in on a very sensitive matter: our ethnic and racial identity. The first essay raises the possibility of a major redefinition of ethnicity, and the second underlines this option of radical change by heralding the arrival on the scene of a type of black individual very likely to make a major difference in the lives of us all, regardless of racial origins.

Part Three focuses on our sacred and also our related, if secular, moralistic options. The first essay explores several major forecasts where formal religion is concerned, while the second updates the Ten Commandments for a New World (ours ... tomorrow).

In Part Four, we lighten the subject a bit, though conceding nothing to the other sections where the importance of making sound choices is concerned. Attention goes to the options apparent in entertainment, recreation, sports, and travel—all of which promise mind-boggling changes likely to pose intriguing challenges and opportunities.

Finally, a capstone essay helps explain how to grab hold of the best ideas from the foregoing, turn them into firm aims, and soon thereafter convert the aims into proud accomplishments.—Editor

Part One

GETTING STARTED—TOMORROW

In the long run, the most important thing about global teenagers is that, like the baby Boom in America, their character will be the main event of world culture. ...
—Kevin Kelly
Whole Earth Review

Essay one tackles the eternal question, *where is there someone for me (that I might be for them)*? How am I to find another with whom to complete myself (a good definition of *love*)? Where am I to find new good buddies? Can the Internet help? And how different will this quest be tomorrow from today?

The second essay explores the role fashion may continue to play in the game of life. Fashion and change are almost synonyms, so much so that "trends seem overplayed right out of the gate, even though the rest of the world is just catching on."[1] In addition, fashion serves many different functions, not all of which are in your face. ("Victoria's Secret catalogs are not inherently subversive, but do we doubt that they advertise something other than underwear?")[2] Accordingly, the essay naturally reaches matters of sex appeal, but it also goes far beyond to discuss, among other novel matters, computerized clothing, including shirts whose built-in electronics measure your vital signs in real time—and can send a health alert to your doctor's computer.

The closing essay carries the message to one logical and popular conclusion, namely, marriage, family formation, and married life (note, however, that longer, and even a permanent state of, singlehood seems on the rise). Bright aids to family life tomorrow are highlighted, especially of the Internet variety.[3] In addition, there is discussion of the key implications of increasing diversity in family types (Only 10 percent of families today are traditional; whereas in 1950, some 81 percent of all households were made up of married couples. Today the figure is 51 percent and falling).[4]

Taken together the three essays raise hope that tomorrow we will have better ways to assuage our loneliness, to give and enjoy love, to dress up as we have always (secretly) wished, and to thrive in the quality of family life we dream about. While there are pitfalls along the way, and Murphy's Law remains operative, looking ahead—with the help of these essays—should help.—Editor

NOTES

1 Stephanie Kang, "Hot Item or Old Hat?" *Wall Street Journal*, 24 October 2003, B-1.
2 Ann Beattie, "First, Let's Kill the Lawyer," *New York Times*, 1 February 2004, 6.
3 Ginia Bellafante, "Cover Baby's Ears: Mommy's Online," *New York Times*, 21 March 2004, 9–12.
4 Faith Popcorn, Quoted in Suzanne Vranica, "Questions for ...," *Wall Street Journal*, 7 January 2004, B-4A.

■ Essay One ■

HEART FUTURES: THE FUTURES OF FRIENDSHIP AND ROMANCE

Marcus Bussey
Teacher/Futurist

Looking out over the next ten years, who will marry or live together with whom? Will your partner share your racial makeup? Does it matter? What is all the fuss about gays marrying? Will pharmaceuticals transform the nature of sexual intimacy? (Is it performance that counts?) Is the risk of disease (AIDS, STD, etc.) changing how we look for love and friendship? Given the proliferation of Internet sites and techno gadgets that offer to identify potential friends and mates, how will you meet people?

Only this much seems clear: Uncertainty and anxiety are the rule. As recently as 1995, Barbara Dafoe Whitehead and David Popenoe of Rutgers University's National Marriage Project, concluded teenagers were awash in choices—and not necessarily happy with this.[1]

"In the final analysis, then, the surveys point up a growing disparity between teens' ideals of marriage and family life—which, if anything, are stronger than ever—and their pessimism about actually achieving those ideals. In part, this disparity may reflect the teens' own life experiences in growing up with divorce, single parenthood and cohabitation. It also surely relates to the general cultural shift toward tolerance of alternative views and choices."[2]

A high schooler in the 1950s had life choices that were relatively simple and clear-cut: marriage, work, a family, home, and the local club. Today, hardly anything is clear-cut.

Instead, a high schooler in the early twenty-first century

may have an insatiable thirst for novelty, one that makes him/her wonder if a perfect mate might be just around the next corner (rather than an arm's length away). This can make commitment difficult and cause restlessness and dissatisfaction. They may have an exaggerated sense of autonomy and individualism. Many may need speed in all they do—and believe it takes too long to build a relationship. Some think they have to live life now—a form of self-centeredness that can and does undermine commitment. These same teenagers, however, may *also* have a comparable need for quality relationships built on acceptance, trust, and loyalty.

LOCAL BECOMES GLOBAL

Thanks to dazzling advances in information technology any of us can find new friends and even romance in cyberspace. That is, we can join networks of like-minded people now defined by interest rather than by geographic space. It is now possible to be friends with someone over many years, and yet not to have met them in person. We may also experience a dose or two of cyberlove with someone we will never meet. We can say "goodbye" to the girl or boy next door and look further afield for that special someone we are "destined" to meet.

Most of us still meet through our networks of friendships and work, but with increased mobility and higher levels of social isolation, people are turning to technology to fill the gap. A fast and fluid lifestyle often results in people deferring marriage. The result is that now in the United States up to one in four homes have a single occupant. This trend to isolation is highly likely to continue.[3]

Networks, extended through the Internet, can help people to reach out and reduce their isolation. Cyberspace offers unlimited opportunities for us to forge intentional communities. Take for example, Friendlyfavors.com. This network offers the opportunity, if you desire, to connect with social activists working for global renewal. Friendships and associations can grow out of a shared passion and value base.

On the other hand, Friendster.com and Friendfinder.com

get down to basics, offering us the chance to overcome our loneliness by creating networks of other like souls looking to broaden their social arena. According to *Time*, one in ten hits on the Internet is for sites like these.[4]

REAL BECOMES MADE UP

Another interesting twist with the Internet is that we can experiment with our identity. In cyberspace, we are all noncorporeal beings capable of presenting ourselves as our fantasies see fit. In "Social Problems of the Future," Lorne Tepperman and James Curtis note:

"Women can experience ungendered interaction for the first time. In turn, this means that people in cyberspace can create new social and sexual relationships. For example, people on line can switch gender and, in this way, disrupt previously held beliefs about gender. Bodily attributes, such as sex or physical attractiveness, are simply irrelevant in cyberspace—at least in principle."[5]

Life and love become a journey of infinite possibility. The downside is that everything, including falling in love, gets ever more complex.

THE HEART WITHOUT BORDERS

The emergence of cyberlove as a force in our society is just another affirmation of humanities' ability to form relationships of substance in unlikely places. Love, like finance, does not seem to respect borders. The Internet has ended the tyranny of distance, communication is instant, and the human ability to take up a space, be it real or virtual, and apply technology to issues of personal happiness appears limitless.

The opportunity before us as a species is to redefine the nature of relationship. Will the old rules (responsibility, commitment, and trust) still apply in a world that can include virtual relationships? For example, are we wrong to suppose we need to touch someone to care? To love? What role does trust play in the sharing of ourselves? How will relationships work in the future?[6]

When looking for intimacy, as was noted previously, cyberspace bestows a certain degree of freedom. We can confess our darkest secrets, our hopes and fears, to someone we may never meet. Is this going to be enough? What remains constant throughout the shifts in technology and culture is the human need for relationship. Isolation is an increasingly tangible part of the modern lifestyle and we can engage with our need for connection by accessing technological avenues. There is nothing phony or foolish in doing so. The Internet is simply a way to augment our social lives.

As technology continues to free us from the constraints of geography and demography, we enter a "brave new world" in which wisdom must become a tool ensuring futures that will genuinely sustain us—heart, body, and soul.

BEYOND OUR COMFORT ZONES

Attitudes are shifting regarding interracial relationships and marriage. People are voting with their feet. Roughly one-and-a-half million American couples today are in an interracial marriage, and while this is not a huge figure (0.6 percent of the total) the figure is the highest ever, and it is doubling every decade.[7]

Furthermore, up to 40 percent of Americans have dated someone of another race, with up to 30 percent of Asians and Hispanics marrying outside their racial groups.[8] Attitudes are definitely becoming more liberal. But tensions are also increasing as the Pew Research Center observed in its "The 2004 Political Landscape" report: "There has been a decided shift across the political spectrum in favor of tolerance in personal attitudes relating to homosexuality and race. Policy matters are a different story, however, and there is considerable evidence that gay marriage, in particular, could open a new chapter in the nation's cultural wars."[9]

Relationships will bear the brunt of these "wars" as it is through our connections, communities, and families that so much of our identities are formed. The heated debates around gay marriage will certainly polarize communities already feeling threatened by the speed of social change.[10] Like abortion and euthanasia, individuals and groups could be caught up in cycles of violence and hate crime.

The race card will continue to attract increasingly populist responses as white Middle America feels the "heat" in an already stressed and increasingly globalized environment. Tolerance is the only alternative to hate and it is the work-in-hand for the coming generation of teenagers. Knowledge of the issues is key to developing open heart futures of quality. The Internet is a conduit of information, and as a route for forming relationships, it promises to be a big player in this future.

EMERGING ISSUES

The Internet, of course, is not the only emergent issue that is reshaping the way we forge relationships. Consider the politics of love. Conservatives have always regarded heterosexuality as the building block of the state and the economy. A shift toward the legalization of same-sex marriages is changing that basic social equation. Many gay couples are seeking the same rights and legal protections before the law as their heterosexual counterparts. Gay-rights lawyers used the American constitutional commitment to equality to argue for same-sex marriages in Massachusetts, and won.

This is a "hot-button" issue that will help define friendship and romance in the coming years. Like the War on Terror it will force each of us to work out our own position, reflect on our ethics and values, and take a stand.[11]

Consider, as well, the economics of love. The family is a stable social unit that generates limited capital. High social mobility, loneliness, and the reckless period of courting generate rapid capital turnover. The culture industries promote, and will continue to promote, a desire for the new, the exciting, and the reckless. Furthermore, there is big money to be made from divorce, which, according to George Draffan of endgame.com, generates $20 billion a year in the United States alone.[12]

This is what the alternative rock group Red Hot Chili Peppers challenge in their hit "Californication." They draw the connection between social control and the economics of instant gratification: "This is what you're craving." Beware a future in which the economics of instant gratification becomes the economics of love.[13]

Consider the chemistry of love. Once, a piece of chocolate was enough to stir pheromones and endorphins, now there are much stronger chemicals available. Viagra, Yohimbe, Damiana, and Androstenone Pheromone, for instance, are popular aphrodisiacs and are readily available from drug stores and online. As the chemistry of love emerges as an offshoot of our study of the human brain, we will undoubtedly see an explosion of erotic drugs in the coming years. As hormone-assisted therapies advance, there will also be ways to modulate gender and the rules for relationship may have to shift as a result.[14]

Consider the wisdom of love. Could schools help prepare young people to be sensitive, caring, and daring lovers?[15] Or should we look instead to a reinvented family, once the source of so much wisdom and strength? Perhaps the media might rethink how relationships of love and trust are portrayed. Can the Internet help us educate ourselves? Should we buy gadgets we program with our interests and use as homing beacons to find what we think we are after?

Finally, what of communicable diseases? Globalization brings with it the risk of pandemic; is there a future for HIV and other sexually transmitted diseases? Will a super virus appear such that the only safe way to be intimate is through the Internet?[16]

Now don't worry. Many of us will chose to enjoy virtual friendships and even have virtual romances. By 2010, we will be able to see their smiling faces over interactive networks. By 2020, we will be able to construct our emotional needs in all respects except the tangible. We will simply profile the physical, intellectual, social, cultural, and economic parameters, and we can have a *perfect* soul mate. We will even be able to turn them off when they get irritating. No more arguments.[17]

"Deep" questions at the heart of forging and sustaining a relationship will stare back at you from the computer screen just as frequently as they do at present from the eyes of your best friend. There is no way around it. What is certain is any resolution will only come when you can stand back and smile, distilling the lessons of the day with some degree of detachment and wisdom.

SUMMARY

As for far-reaching forecasts, then, these will have to suffice: Your generation will—like everyone before it—wrestle with a fear of love. A fear of stretching boundaries, a fear of intimacy, and a fear of vulnerability. Slowly, and at cost, it will find its own way to love, to making a commitment. It will come to understand the internal logic of love—that we need to give unconditionally and we all desire to be accepted. A romantic novelist put it quite well: "Real romance, ... may well be a bunch of flowers or a loving kiss when it's least expected, but it is also when someone does something for another that is slightly beyond the call of duty, when we stretch our boundaries, to become, just for a moment, a truly romantic human being."[18]

NOTES

1 BarbaraWhitehead and David Popenoe report that, "Marriage is viewed as an important part of the future for most high school seniors. They say they most likely will choose getting married in the long run, and this attitude has increased slightly over the years from 77 percent in 1975 to 80 percent in 1995 (Monitoring the Future). In 1995, only 4 percent of the seniors answered that they probably would not get married, with the remaining 16 percent either already married or answering that they "have no idea."

A 1992 Gallup Poll of youth aged 13 to 17 years found an even larger percentage who thought they would marry someday—88 percent compared with 9 percent who expected to stay single. Gallup has undertaken a youth poll several times since 1977 and the proportion of youth expecting to marry someday has not varied much through the years (Bezilla, 1993).

The great majority of high school seniors also strongly believe in the importance of marriage and family life, but more so for girls than for boys. Between 1980 and 1995, there was an increase from 75 to 78 percent in the proportion of seniors who said that having a good marriage and family life were "extremely important" (another 14 percent answered "quite important"). The proportion of girls who agree with this statement is consistently 9 to 12 percent higher than the proportion of boys (Monitoring the Future). In a recent survey of the nation's college freshmen, however, there was not much difference between the numbers of males (70 percent) and females (72 percent) who consider raising a family to be an "essential" or "very important" objective (The American Freshman)."
http://marriage.rutgers.edu/Publications/Print/Print%20Teen%20Attitudes.htm.
You should note that the Marriage Project has its critics, who claim its values orient marriage as the desired end of relationships. See: What's Wrong with the Work of the National Marriage Project?: http://www.unmarried.org/10problems.html.

2 Ibid.
3 Kerby Anderson, "Loneliness": www.probe.org/index.html. Loneliness also links with depression and is part of the profile of depressed individuals.
4 Lisa Takeuchi Cullen, "Cupid Academy," *Time*, 16 February 2004, 60.
5 Lorne Tepperman and James Curtis, "Social Problems of the Future," *Journal of Futures Studies*, vol. 8, no. 1, 2003, 31.
6 The internet is a fascinating place for those of us who like to explore virtual reality, but it is not for everyone. Social class membership, gender, economic conditions, social aspirations, and the very nature of people shape their choices when it comes to looking for friendship and love. (*Time* recently polled a group of volunteers in the 2004 democratic presidential preselection campaign. The staffers for Howard Dean, Wesley Clark, and Joe Lieberman all used the dating Web site Friendster.com. Those working for Richard Gephardt, John Kerry, John Edwards, and Dennis Kucinich did not. Is there a pattern here)? Joel Stein and Romesh Ratnesar "New Hampshire: Scoping Out the Volunteer," *Time*, 26 January 2004, 21.

7 See: http://racerelations.about.com/library/weekly/aa020903.htm.
8 Ibid.
9 See Pew Research Center for the People and the Press at www.people-press.org. See also "The 2004 Political Landscape" at www.people-press.org, 4.
10 See *Time*'s "The Battle over Gay Marriage" by John Cloud, 6 February 2004. This article gives a balanced account of the debate to date and offers interesting statistics based on party preference and also a historical overview with timeline of the gay marriage movement.
11 If you are Christian and feeling confused, there is an excellent text by L.R. Holben called *What Christians Think About Homosexuality* (Berkeley, Calif.: Bibal Press, 1999). It and other thoughtful texts are reviewed at http://www.someone-to-talk-to.net/new_page_16.htm. The unnamed reviewer had this to say about Holben's book: "This is an important foundational book and I suggest that it be one of the first books you read. The author summarizes six representative viewpoints that cover the spectrum of current Christian thought on homosexuality. He does not try to convince you that any one viewpoint is correct, but presents the arguments for each in an unbiased manner, shows how it has been criticized, and explains how the proponents of that viewpoint answer their critics. The author's stated goal is to 'synthesize and popularize' the extensive scholarly and theological work that has been done on the question of Christian faith and the homosexual person and to render the complexity and breadth of that material accessible to others who, for pastoral or personal reasons, seek to move beyond polemics and absolutist pronouncements to a thoughtful consideration of the scope of Christian thinking on this issue."
12 George Draffan at http://www.endgame.org/endgame.html.
13 Forgive me if I state the obvious, but humanity has always been fascinated with sex and all sorts of so-called aphrodisiacs have been developed (rhino horn comes to mind for obvious reasons). "Born and raised by those who praise Control of population" The full text is to be found at http://www.lyricsfreak.com/r/red-hot-chili-peppers/114735.html.
14 To get an overview of current research, warts and all, see

Dr. Simon LeVay's 2003 article "The Biology of Sexual Orientation," which can be found at http://members.aol.com/slevay/page22.html.

15 Marriage and relationship education is examined in the following report by Marline Pearson: "Can Kids Get Smart About Marriage? A Veteran Teacher Reviews Some Leading Marriage and Relationship Education Programs" found at http://marriage.rutgers.edu/Publications/pubCanKids.htm.

16 The World Health Organization (WHO), which monitors conditions pertaining to potential pandemics, covers this issue in brief at the following site: http://www.who.int/csr/disease/influenza/pandemic/en/.

17 If you are desperate, get a cat or dog if you don't feel you are fully accepted and understood. Pets take you as you are, in all your vulnerability. So should your friends and partners. Remember that for all the hype, progress—both technological and social—is having very little effect on the human need for companionship. To see how effective pets are in relieving stress, see the short article by Cynthia Chandler, "Animal-Assisted Therapy in Counseling and School Settings," which can be found at http://www.ericfacility.net/databases/ERIC_Digests/ed459404.html.

18 Candida Baker, "The Myth of Romantic Love," in *Loves: From the Sacred to the Profane*, Edited by Jean Bedford (Sydney, Australia: Angus and Robertson, 1995), 13.

REFERENCES

Anderson, Kerby. "Loneliness": www.probe.org/index.html.

Baker, Candida. "The Myth of Romantic Love" in Bedford, J. (ed.). *Loves: From the Sacred to the Profane*. Sydney, Australia: Angus and Robertson, 1995.

Bezilla, Robert (ed.). *America's Youth in the 1990s*. Princeton, N.J.: The George H. Gallup International Institute, 1993.

Chandler, Cynthia. "Animal-Assisted Therapy in Counseling and School Settings": http://www.ericfacility.net/databases/ERIC_Digests/ed459404.html.

Endgame: http://www.endgame.org/endgame.html.
Interracial Relationships:
http://racerelations.about.com/library/weekly/aa020903.htm.
LeVay, Simon. "The Biology of Sexual Orientation": http://members.aol.com/slevay/page22.html.
Pearson, Marline. "Can Kids Get Smart About Marriage? A Veteran Teacher Reviews Some Leading Marriage and Relationship Education Programs": http://marriage.rutgers.edu/Publications/pubCanKids.htm.
Pew Research Center for the People and the Press: www.people-press.org.
Tepperman, Lorne, and James Curtis. "Social Problems of the Future," in *Journal of Futures Studies*. Vol. 8, No. 1, 2003.
What's Wrong with the Work of The National Marriage Project?: http://www.unmarried.org/10problems.html.
Whitehead, Barbara Dafoe, and David Popenoe. "Changes in Teen Attitudes toward Marriage, Cohabitation and Children 1975–1995": http://marriage.rutgers.edu/Publications/Print/Print%20Teen%20Attitudes.htm.
World Health Organization: http://www.who.int/csr/disease/influenza/pandemic/en/.

■ Essay Two ■

THE FUTURE OF FASHION

Linda Brown
Futures/Market Researcher

Did you know it was soldiers returning from World War II who first made T-shirts (Miller, 1999) and flip-flop sandals (Tenner, 2003) so popular in the United States? Robes and scarves will probably be among the elements of comfort and protection that young soldiers returning from Iraq and Afghanistan bring back this time. What is your opinion of Middle East governments requiring women to wear traditional garb? Or the French government's new effort to prohibit wearing headscarves and other religious symbols? Do you believe people may someday wear a special glove to generate energy or protect themselves?

How might we dress in the near future? Why, and so what? Do you find the clothes or costumes portrayed in the future daringly different or downright dull? My five forecasts below may amuse, surprise, or perhaps even turn you off—but they are unlikely to bore you.

WEARABLE COMPUTING

The most exhilarating development in the clothing industry during the next decade will be computerized clothing. Integrating technology directly into textiles and garments is a new trend. "Smart clothing" and wearable applications exist at present, though hardly in the variety and quantities expected to appeal to you long before 2010.

"Wearable computing," like many technologies that eventually change the future, first appeared as science fiction, e.g., the film *Terminator* in 1984 promoted influential images of computer-integrated humans. When it actually made a real-world debut, it looked like the futuristic silver cat suit worn by Tomb Raider Laura Croft in *The Cradle of Life* (2003).

Today, most "wearable computing" fits into a pocket or straps onto the body. It is being developed and used to meet the needs of very specific careers. Early in 2001, for example, Cisco Systems produced iTrack—a wearable computing system that allowed people on the move to make phone calls, send and receive e-mail, stream live video, and log on the Internet—using off-the-shelf technologies. The company anticipated then that soldiers, firefighters, police, and paramedics might wear it in the not-too-distant future (Wardill, 2001).

Similarly, the Department of Computer and Information Science at the University of Oregon is designing and experimenting with NETMAN, a wearable computer system that enables technicians to collaborate between field- and office-based locations. They expect the system to be useful to teams working in maintenance, construction, and manufacturing industries (Kortuem et. al., 1999). Networking teams of workers, athletes, or others should increase productivity and improve performance in many situations. It is also hoped that the development of collaboration technologies may reverse the isolation and alienation that some people have experienced with the rise of the Internet and use of high-tech systems.

The Wearable Internet Appliance WIA-100NB, offered by Hitachi in Japan and by its developer Xybernaut in the United States, brings us closer to the future with laser optics floating a virtual notebook-sized screen in front of you (Smith, 2002). EShades by Inviso is an example of head-mounted computer displays. Wrapping around like a pair of oversized sunglasses, EShades allow mobile professionals to privately view their screen (So Cool, 2000).

Steve Mann, a professor at the University of Toronto, gives us a glimpse into future possibilities. Wearing a computer helmet equipped with video cameras, he alters the reality he experiences by instructing the screen to block out billboards or other objects he does not care to see (Jennings, 2003).

The progress of wearable computing hinges on miniaturizing power systems. Battery units are often bulky, heavy, and inconvenient. Researchers are busy developing fabrics with cells that generate solar electricity and fabrics capable of generating power as they flex or have wind move across them. Biomechanical energy generated by body movements will likely play an important role. The human body is a storehouse of energy. Breathing, blood pressure, body heat, and limb and finger motions are all potential sources of energy. Breakthroughs in power generating and energy storage systems will be decisive in speeding or slowing the future of wearable computing.

MARKET AIM

Targeting specialized markets is a savvy cost-effective way to introduce wearable computing. Today, healthcare and sportswear are the focus. Tomorrow, teens and young professionals who make up the hi-tech fashion field will be the targets. Imagine the variety of applications that built-in monitoring systems might have in any setting.

The Smart Shirt developed at Georgia Tech University is a prime example. Researchers wove plastic, optical, electrical, and other specialty fibers into fabric to produce a shirt embedded with sensors able to monitor breathing. Heart rate and body temperature are also monitored with sensors attached to the body or integrated into the shirt. Financed by the U.S. Navy in 1996, the Smart Shirt was originally designed to identify the location of bullet wounds suffered by soldiers in combat. In June 2000, a Manhattan-based company, SensaTex Inc., made a licensing agreement with Georgia Tech to market it first for medical purposes, with plans to use it in athletic apparel.

French laboratories have also developed a medical T-shirt, linked to mobile communication networks that can signal for emergency assistance if needed, allowing elderly healthcare patients to remain independent while "hospitalized" at home. Researchers with Philips in Germany have discreetly incorporated monitoring, alarm, and storage devices into bras, briefs, and waist belts.

Expanding on the concept of biosensing underwear, the U.S. Army recently displayed a jacket at the "Wear Me" exhibition in London with a custom-built miniaturized air-conditioning system that will allow soldiers to fight and survive in extreme heat conditions. As the classic futures novel *Ecotopia* (1973) suggested, people might soon wear clothing that changes with the weather. We may need such clothing if climates change significantly.

One day you may use a smart shirt to protect a susceptible newborn from Sudden Infant Death Syndrome (SIDS). Training as a college or professional athlete, you might don a smart shirt to improve your performance. Or maybe you will participate in the safety surveillance of a scuba team diving to explore ocean depths, rescue people, or hunt for treasure. Now that clothes with special monitoring and signaling capabilities are available for medical, sports, and military uses, the technologies will be modified for other applications. Dating, for instance, might be "simplified" in the twenty-first century by programming garments or accessories to broadcast the wearer's profile and their preferences in a partner.

If you wear a monitoring or transmission device into a party or on a crowded street, how would your natural inclination to trust or interact with other people change? Today, researchers are studying how mobile information devices help youngsters become actively independent in urban and rural places. Entertaining choices are expected to appeal to youthful markets. For instance, researchers at Massachusetts Institute of Technology (MIT) have "wired" a musical jacket. When your parents yell at you to turn your jacket down, they may no longer mean the collar.

ETIQUETTE AND TRENDSETTERS

The clothes of the twenty-first century require and express new etiquette. Laws may attempt to restrict what can be worn. By regulating how MP3 players may be worn, defining when a cell phone may be permitted, or banning buttonhole cameras from campus, high school dress codes are attempting to cope with the earliest fashionable technologies. School staffers want to preserve a learning environment, respect privacy, and ensure safety. However, creating rules for wearable computing will be far easier than enforcing them.

Trendsetters will continue to trigger new styles tomorrow. Because it is a period of transition, our society in the new century will be tolerant of variations, usually extolling them as evidence of diversity and freedom. At other times, variations will be condemned as purporting role confusion or depravity. Scandalous scenes will stir controversy and fashion, just as the tennis star Gorgeous Gussy, flashing lace from beneath her skirt (ask your great-grandparents), and pop star Janet Jackson, flashing skin at the Super Bowl (don't ask), exceeded the limits of morality in their day. Entering the twenty-first century, today's dual hemlines—short inside and long outside—express the turmoil between public and private values.

Freethinking styles will increasingly originate at the fashion edge, from alternative media, street-styles, or exclusive clubs, where experimentation or racy distractions will be welcome. Extremes could appear in response to a sense of invaded privacy; both flamboyant showmanship and cover-up styles. Which styles move from "freak to chic" (Harris, 2002) will depend upon how Americans perceive themselves. But if young people are drawn rapidly into the workforce to fill labor shortages and fund social security, their trendy styles could move through quickly unless the boss demands serious, conservative attire.

Replacing fashion's use since the 1960s as a fulcrum for equality, status dressing is returning, marked more by quality and occasions, and less by labels. The rich use of fabrics will still proclaim prosperity for some, but in times of uncertainty, success will be modestly understated or even hidden. Increasingly, those who can afford it will invest more in enhancing the body beneath their clothes with life-sustaining health or surgical perfection.

Politics set trends too. Do you need any more reason to vote? Believe it or not, the president influences what is or isn't cool to wear. So does war. Once-dominant duds can fade in popularity if a war makes them unpopular. Rationing is also a possibility during times of war.

Exerting authority, in 1909 the French instituted a regulation that forbid women to wear masculine clothing except when riding a horse or a bicycle. It took more than 120 years—six generations—for trousers to become truly acceptable for women to wear. The full acceptance of skirts for men should come much sooner, because they have a historical and cultural basis, and the metrosexual male is being marketed.

Twenty years ago, when the roles of women in the workforce and the importance of power dressing were the talk of the town, most Americans interpreted men wearing skirts in our society as cross-dressers. Today's generation is more accepting, having been exposed to bold, burly, or striking costumes of media icons—the leather-kilted *Gladiator* Russell Crowe, the resourceful pirate-like Rupert of *Survivor*, the No Doubt rock star Adrian Young, and the floor-length dusters worn in *The Matrix*. In certain situations, a full, wrap, or buttoned-loin skirt adds a layer of protection or privacy to be appreciated by either gender.

To celebrate the opening of *Braveheart: Men in Skirts*, nearly four hundred men in kilts marched through Manhattan in 2004 to the Metropolitan Museum of Art (Rayner, 2004). Now that designers are reintroducing lace and floral prints into men's clothing, why should something like the Workman's Utilikit—a skirt that is essentially a glorified tool belt—be considered unmanly? To carry the tools of this century, sexless aprons, pocket-laden vests or jackets, and satchels will be sensibly accepted. Workplace opposition to items once considered effeminate will ease, and looking out to 2010, men should steadily regain more fashion freedom.

Have you noticed that unisex styles show up most often in sports and professional settings? Sameness is accepted to express team unity and to encourage gender equality. You will see more "unit dressing" as families, tribes, believers, and other groups express that they are united in purpose. Unique expressions of individualism will be better tolerated wherever competition gives way to collaboration. Opportunities to wear sensuous clothing will never go away completely. Instead, new developments are compelling consumers to make greater distinctions between genders.

Color coding girls in pink and boys in blue did not begin until the mid-twentieth century so, in the interest of gender neutrality, it would probably be easy to give up this short-lived trend. However, because prenatal screening and genetic engineering allows parents to know or even control the gender of their babies before birth, layettes for newborns are more likely than ever to be blue or pink instead of a once-popular gender-neutral yellow or green.

Families have fewer children too. They can spend more on the few they do have and worry less about needing neutral styles to pass among several children. Dressing girls in blue has been trendy while reformers try to argue that girls can do anything boys can do, but it also implies "blue is better." Now that gender can be genetically selected, should the new message be "pink is equally perfect?" If people should be genetically engineered or cloned, what color clothes will a child wear if it was planned and designed to be androgynous?

EMPHASIS ON FORM AND FUNCTION

The shapes of things to come will also be determined by their suitability for our surroundings. Concerns over comfort and convention ultimately give way to basic needs.

Declining birth rates have caused some European countries to offer financial incentives for women to bear and raise children. As long as fertility, childbearing, and childrearing are valued, styles will cling to displaying femininity. Because an hourglass figure is a sign of fertility (Etcoff, 1999)—with or without the aid of corsets—curvaceous cuts will continue to reappear in women's wardrobes.

The return of tailoring (Woolnough, 2004) will also accentuate the human form. The sacky shape of easy-on easy-off clothes could be associated negatively with aging. Propriety will simply limit how and where figure-flattering garments are worn by either sex, but clothes will never be completely unisex. They will always be ruffled by romance, especially when the benefits of being female are appreciated, unless we evolve to some new species or produce most offspring in labs.

Labs are producing new fabrics, some made from recycled materials. Recently developed, Crabyon is made from the previously discarded parts of shellfish and possesses remarkable antibacterial, antiodor properties (Roberts, 2003). Fabrics like this may put an end to the offensive smells in your schools, gymnasiums, and locker rooms. Wouldn't it be great if, instead of your ingesting pills or enduring shots, antibacterial clothes could ward off illnesses?

After the U.S. government issued terrorism alerts, demand for survival gear increased fivefold in February 2003 (Selingo, 2003). It would be difficult to live a normal life wearing or carrying gas masks. Hopefully it will never come to that. But outer garments could increasingly be made of chemical-resistant materials. Cotton, leather, and canvas are all absorbent materials, and therefore may decline in popularity as an outer shell. Consideration must be given to dressing safely, probably in layers. If you carry hand-cleaning gel in your lunch kit, have used a bandanna to avoid breathing dust, or have pulled the knuckle-long sleeves of your trendy top over your hand to handle an object, you have already thought about taking some precautions.

Practical concerns about the environment could force the use of the safest fabrics, colors, and styles. Some precautions will be sensibly accepted, like black clothes in Britain's sooty coal-powered cities of the nineteenth century. Others, like ending the early twentieth-century fetish for feathers, will require acts of legislation, taxes, or fines instituted in response to persistent lobbying.

Imagine the gown that may be worn at your wedding. If you imagined it being an ostentatious white, think again. A gown that could be refashioned into a bed spread, other linens, or daily wear could represent unending commitment. In the interest of a sustainable environment, designs will be expected to endure. Americans—many once accustomed to excessive amounts of clothes—may find less is better.

Just remember, if grandma offers you her old sewing machine, take it. You might need it to tack in an attractive tighter-fitting form. You could easily find a use for it crafting a project of recyclable materials. Having the skills needed to take part in the creation of tomorrow's fascinating fashions could be a great benefit too.

TECHNOLOGY REVOLUTION

Fashion manufacturing and retailing are being transformed by technologies. Expect your shopping experience to be considerably different tomorrow than today. At present, the vast majority (80 percent) of clothing worn in industrialized countries is not "fashion" at all (Clancy, 1996). True fashion displays the "art of high-sewing." Only about three thousand women a year throughout the world buy made-to-measure fashion (Baudot, 1999). Tomorrow, more upper- and middle-class consumers could have access to truer "fashion" that may or may not be created by a famed designer or derived from one of the twenty-one major fashion houses.

Bringing made-to-measure clothes to those well-heeled middle- and upper-class individuals would be a significant but feasible change. After sipping your favorite latté at the local coffeehouse, pop over to the storefront next door so your body-wrap measurements can be taken for customized clothing. Maybe the shop owner will be a cousin of Joseph, the tailor who already markets over the Internet and then travels to the United States to custom fit clothes for his clients. Should it matter where the clothes are made?

Paris, London, New York, and Milan are traditional fashion centers, but influence is known to wax and wane. The randomness of the Internet introduces opportunities for ideas to come from most anywhere and be shared with most anyone. If entrepreneurs could break into fashion before the Internet, how much greater are the opportunities for creative talent today?

Before the turn of the century, even famed designers shifted their seasonal presentations to CD-ROM collections and have now moved to Web site displays instead of presenting a runway show. IBM has since hosted a fashion show in Kuala Lumpur, Malaysia, where students displayed the interconnections of Information and Communication Technology (ICT) and clothing (Ismail, 2003). Deja View Inc. hopes to market its lipstick-size wearable camera through major retailers that specialize in electrical products, such as Circuit City and Best Buy (Willis, 2003).

If wearable computing were sold at your local Radio Shack, traditional retailers would take a hit comparable to the 1970s' switch to selling sportswear in sporting goods stores. Wherever they are sold, watch how embroidery, cording, snaps, and metallic fabrics appear in our clothes. It would be a safe bet to say fashions will be "snappy" (no pun necessarily intended), because circuits can be formed when the fabric is pierced with metal fasteners, still allowing devices to be removed for cleaning. How good is the dry-cleaning service you use now?

To learn about wearable computing, your time could be better spent now at a healthcare, computer, or media industry conference instead of a traditional fashion show. Computers are changing where fashion is created, who designs it, and how it is marketed, as well as what is worn. As the Father of Haute Couture, Charles Worth, elevated the status of lowly dressmakers by making some highly valued artisans, this current technological revolution could alter the status of those involved in the manufacturing of clothes. Who will be a champion this time?

College students across the nation are among those advocating change this time. United Students Against Sweatshops (USAS) is active on 175 college campuses (Bolino, 2003). January 1, 2005, marks a pivotal day for U.S. textile workers, and those in other industries as well. On which side of that date do you stand? Did you realize a crucial day was passing? Under agreements of the World Trade Organization, U.S. markets are scheduled that day to continue opening their markets to imports from countries that currently have lower production costs. From where you stand, have workers and the environment also gained protections?

The Bureau of Labor Statistics has forecast declining employment for apparel workers in the United States through 2010 (BLS, 2002–2003). Others believe the labor market could potentially be reorganized as a consequence of the Internet economy, if the value of human knowledge, skill, and labor is defined (Davis & Meyer, 2000). Would Americans accept such a tiered class structure?

A reversal in transportation trends could also alter the expected employment outlook. International shipping costs have steadily decreased since the 1960s, but increasing security restrictions or fears about the safety of oversea products could renew interest in local production.

In past centuries, fashion designers often studied art or retailing. The most rewarding fashion careers of our century will require an understanding of international trade, computer technologies, and material sciences, as well as an appreciation for the flow of fabrics on a person's form.

SUMMARY

The flux of fashion through time makes a wonderful study. You can expect people will continue trying out the styles of cultures they encounter, embracing some fully. As before, we will embellish our clothes and bodies as permitted with symbols of what we admire. We will continue to experiment with new materials and designs, in reaction to the exploration of inner and outer worlds, especially space, the human genome, and consciousness as we enter new eras.

To anticipate the future of fashion, you do not have to work in a hi-tech laboratory, design studio, or think tank. Just pay creative attention to the world around you—and allow yourself to enjoy the sights and sounds of our wonderful world of garb.

REFERENCES

Baudot, François. *Fashion: The Twentieth Century*. New York: Universe Publishing, 1999.

Bolino, Monika. *Fashion*. San Diego, Calif.: Greenhaven Press, 2003.

Bosner, Kevin. How computerized clothing will work. Howstuffworks (on-line): http://computer.howstuffworks.com/computer-clothing1.htm (accessed January 2004).

Breward, Christopher. *Fashion*. Oxford U.K.: Oxford University Press, 2003.

Bureau of Labor Statistics (BLS), U.S. Department of Labor, Occupational Outlook Handbook, 2002–03 edition, Textile, Apparel, and Furnishings Occupations, on the Internet at http://www.bls.gov/oco/ocos233.htm (accessed 15 January 2004).

Clancy, Deirdre. *Costume since 1945: Couture, Street Style, and Anti-Fashion*. New York: Drama Publishers, 1996.

Cristol, Hope. "The Future of Wearable Computers: From Assisting the Disabled to Serving as Secretaries, Wearable Technologies Will See Their Practical Uses Expand." *The Futurist*, September–October 2002, 68–69.

Davis, Stan, and Christopher Meyer. *Futurewealth*. Boston, Mass.: Harvard Business School Press, 2000.

Etcoff, Nancy. *Survival of the Prettiest: The Science of Beauty*. New York: Doubleday, 1999.

Fernie, John, Suzanne Fernie, and Christopher Moore. *Principles of Retailing*. Oxford, U.K.: Butterworth Heinemann, 2003.

Golbin, Pamela. *Fashion Designers*. New York: Watson-Guptill Publications, 1999.

Harris, S. Dyer. "Where Is Your Next Big Idea Coming From?" *The Futurist*, November–December 2002, 21.

Holsendolph, Ernest. "Georgia Tech Lands Licensing Agreement for Medical Garment." *Atlanta Journal-Constitution*, 14 June 2000.

Ismail, Izwan. "Tech and Trendy." *New Straits Times*, 9 October 2003.

Jackson, Leigh-Ann. "Susan Dell to Close Austin, Texas, Clothing Boutique, Shift to Direct Sales." *Austin American-Statesman*, 26 August 2003.

Jennings, Lane. "From Virtual Communities to Smart Mobs: Wearable Computers and Phones Offer a Transparent Future. Should We Trust It?" *The Futurist*, May–June 2003, 6–8.

Kidman, Angus. "Off the Cuff: Wearable Computers Still Have a Way to Go." *Australian*, 17 June 2003, T09.

Kortuem, Gerd, Martin Bauer, and Zary Segall. "NETMAN: The Design of a Collaborative Wearable Computer System." *Mobile Networks and Applications* 4, 1999, 49–58.

Mathews, Ryan, and Watts Wacker. "The Deviant's Advantage: How Fringe Ideas Create Mass Markets." Crown Business Publications: www.randomhouse.com/crown/business, 2002.

Miller, Brandon Marie. *Dressed for the Occasion: What Americans Wore 1620–1970*. Minneapolis, Minn.: Lerner Publications Company, 1999.

Orth, Maggie, and E. Rehmi Post. *Smart Fabric, or Washable Computing*. Cambridge, Mass.: MIT Media Laboratory, 1996.

Philips Research. "Biosensors for Underwear." *The Futurist*, January–February 2004, 13.

Quinn, Stephen. "News Tools of the Robo-Reporter." *Australian*, 27 April 2000, M14.

Rayner, Abigail. "Bravehearts Who Refuse to Toe the Hemline." *Times* (London) *Overseas*, 7 February 2004, 24.

Roberts, Sonia. "Food Waste into Fragrant Fashion." http://just-food.com/news (accessed 2 October 2003).

Selingo, Jeffrey. "For Some, the Jitters Help the Bottom Line." *New York Times*, 20 February 2003.

Smith, Tony. "Wearable Computers Fit for the Catwalk." *Times* (London), 5 March 2002, BT31.

"So Cool You Need E-Shades." *Minneapolis (Minn.) Star Tribune*, 11 October 2000, 13E.

Tenner, Edward. "Our Own Devices: The Past and Future of Body Technology." New York: Alfred A. Knopf, 2003.

University of Bath, Public Relations Office. "Wearable Power." *The Futurist*, September–October 2003.

Wardill, Steven. "Technology Keeps Wearers on Track." *Brisbane (Australia) Courier-Mail*, 6.

"What's the Word? Dress Code in Wearable Computing." *London Observer*, 11 January 2004, 12.

Willis, David. "Company to Debut Wearable Camera." *Asbury Park (New Jersey) Press*, 7 August 2003: http://www.app.com

Woolnough, Damien. "It's a Man's World." *Australian*, Features, 30 January 2004, 16.

FURTHER READING

Ewing, Elizabeth. *Everyday Dress, 1650–1900.* New York: Chelsea House Publishers, 1984.

Gardner, Robert, and Dennis Shortelle. *The Future and the Past.* Englewood Cliffs, N.J.: Julian Messner, 1989.

Irons, Diane. "Teen Beauty Secrets: Fresh, Simple & Sassy Tips for Your Perfect Look." Sourcebooks, 2002.

McHugh, Josh. "Wrist-Top Revolution." *WIRED*, April 2003, 96.

Rifkind, Hugo. "Trousers Are Just So 1800s." *Times* (London), *Overseas*, 7 February 2004, 24.

WEB SITES

Bristol Wearable Computing project:
http://wearables.cs.bris.ac.uk/

Fashion Schools.org:
http://www.fashion-schools.org/

Fibre2Fashion, World of Garment–Textile-Fashion:
www.fibre2fashion.com/

Fifth International Symposium on Wearable Computers:
http://iswc.gatech.edu/

History of Fashion:
http://alpha.furman.edu/~kgossman/history/

MARS—Mobile Augmented Reality Systems:
www.cs.columbia.edu/graphics/projects/mars/mars.html

The MIThril at MIT Media Lab:
http://www.media.mit.edu/wearables/mithril

Mytailor:
www.mytailor.com/

National Association of Schools of Art and Design (Reston, Virginia):
http://nasad.arts-accredit.org/index.jsp

Steve Mann, Professor, University of Toronto:
http://wearcam.org/

Wearable Computing:
www.wearablecomputing.com

Xybernaut:
www.xybernaut.com

■ Essay Three ■

ROMANCING THE CLONE: THE FUTURE OF THE FAMILY AND RELATED ISSUES

Marilyn Dudley-Rowley, Ph.D.
Lecturer, Sociology Department, Sonoma State University

Looking out toward 2010, what sorts of changes are likely in the American family? I have ten forecasts to share, and expect at least a few of them to raise your eyebrows. Changes in family life over time have *always* created surprise, and either applause or disdain, depending on age, gender, and the other major ways we have invented for sorting one another.

HIGH TECH TO THE RESCUE

To begin with, we will probably find mates in crisp new ways, at least as compared to how our folks managed it. Now, we can "hook up" through on-line dating services tailored to whom we want to interact with. Expansion of the Internet, along with fast transportation and other quick reliable communication, has transnationalized sex, marriage, and family life.

I am a living example: Married to an Argentine man who works in Russia as a cosmonaut, I am the stepmother of two little boys who live in Moscow as a result of my husband's first marriage. I would have never met him had there not been the necessity for us to initially work together on a space project using e-mail.

We can now instantly Google our way to eligible partners on the other side of the planet and even communicate in their languages using the translation tools of the Internet. Doubtless, many teenagers reading these words have already met someone about their own age on-line and can discuss some of the pitfalls of e-mail or chatroom friendships.

In short, it is *computer* technology, the Internet, and their globalizing effects that are creating most of the impacts on the family and related issues. Computer technology, which makes possible rapid exchange and processing of information, can increase the frequency of human interactions at an exponential rate. The speed of social change is partly a function of the speed and ease of these interactions.

TAKING MALES OFF THE PEDESTAL

A second forecast concerns *patriarchy*, a system by which males dominate in most aspects of society. It is being graded down to help America achieve a more level playing field (and give females equal opportunities). This change has more to do with the information process described above than it does with any one landmark event or important piece of legislation like USC Title VII.[1] As gender steadily loses importance, things that relate to it—such as marriage, family, sex, and reproduction—will quickly change in response.

It is logical, for example, to believe that a weakening of patriarchy would greatly expand the role of fathering, especially as more and more households shift toward dual-earner status. Naturally, this has not happened with the speed of light, given social and cultural *drag* from the days of breadwinner dads who had a very small role as fathers. That notwithstanding, I forecast parenting by fathers will increase, especially as the number of children per household is falling. (Some of you reading this may know a teenage father. How have the duties of fatherhood complicated, and, hopefully, also enriched his life?)

GAY UNIONS

A third forecast focuses on the unexpectedly promising future of same-sex relationships. Without the legal protections of marriage, gay and lesbian partners often find themselves out in the cold in contracts pertaining to their mutual property rights, inheritance, and responsibility to each other. To give an example, many a homosexual partner has been barred from the bedside of a sick or dying partner in the hospital by parents who opposed the partnership from the outset. San Francisco Mayor Gavin Newsom has argued forcefully, "Same sex marriage is the civil rights issue of the twenty-first century."[2] (In California, organizations for gay teens are active in our high schools. Is it the same at your school? Why or why not?)

MORE FROM MORE

The rights of same-sex couples will not be the only challenge to the man-woman definition of marriage. Though you might at first think I am joking, the issues of same-sex marriage and different cultures will bring an old practice into new light.

Polygamous marriage, typically between a man with two or more wives, is a legal fact of life in more than half of the world.[3] During the 1960s, people experimented with different configurations of conjugal arrangements, living communally with many partners or openly having many partners outside of the household. The *open* marriage concept where spouses "held each other with open arms" was a popular idea in the 1970s. These attempts to address polygamy in the American context were a serious challenge to monogamy.

Same-sex marriage will open a great many legal doors, blocking legal polygamous unions. Polygamy will not be popular with those who have strong religious, philosophical, or personal reasons for desiring monogamy.[4] However, it will be popular among those who desire variety in companionship and the economic benefits of a number of males and females living together. Such communal marriages might not quite be the epitome of the "heart without borders" as Marcus Bussey writes about in "Heart Futures" in essay one of this volume, but it certainly calls for those with more expansive hearts.

TRY, TRY AGAIN

The growing economic independence of women, and longer life spans, make the "'till death do us part" notion untenable in practice. *Serial monogamy* (one-after-another marriages dissolved by divorce) is likely to prove an evermore common fact of American life.

This increase will lead to an upswing in *blended families*. Such families, especially those that retain some semblance of cordial relations among members, pose almost as big a challenge to the system of laws in advanced industrial societies as do same-sex marriage and polygamy. (Many teens reading this piece may live in blended families and can undoubtedly relate to the ups and downs of having stepparents and stepsiblings.)[5]

WHY MARRY AT ALL?

Cohabitation, a couple living together as husband and wife, often temporarily until marriage, is on the upswing, due in large part to the relative economic independence of women and their need to take the time—into their 20s and 30s—to establish a career. While likely to expand, it does not pose the same legal challenges as the developments cited above. Many states already use English common law, as in South Carolina, where a common-law union *is* a legal marriage that requires a formal divorce to dissolve.

BREAKING UP? CLICK HERE

Obtaining marriage licenses and enacting relatively uncomplicated prenuptial, postnuptial, and no-fault divorces are currently within the grasp of our computer-based information technology. Uncomplicated child-custody and property-division agreements would not be much further off after these procedures become common.

If the notion of marriage is changing, so is that of divorce.[6] Already several on-line services are helping couples split up faster and cheaper than through traditional legal routes. With the role of the state becoming more prominent in monitoring parenting, children divorcing their parents may become more common, but it won't be as easy as "point and click."

SELF-HELP GAINS

People will increasingly avail themselves of self-help from popular literature (e.g., "how-to" books, newspaper columns, and magazine articles).[7] They will turn to an emerging counseling medium on TV (Dr. Phil, etc.) and on the Internet. They will escape bumpy marriages (perhaps with the click of a mouse). Some will turn to "lifestyle drugs" like Viagra to help with sexual performance. Provided they exercise appropriate caution, and look to one another for honest feedback, most should do quite well.

CHILDBIRTH CHANGES

As with marriage, women have been putting off having children until increasingly later in life. Currently, childbirth is not recommended for women past the age of thirty-five—for complications to the health of mother and fetus multiply after that age. However, even the biology and technology of reproduction finds itself at the beck and call of social forces. So, we may soon see large aggregates of women in their forties and older becoming new mothers.

CLONING, ANYONE?

Complications and fertility issues in these later years will merge with technological advances in *cloning* (the reproduction of a complex organism from its own cellular material). In time, we will merge what we learn from the technology of cloning with sophisticated incubators that, for all intents and purposes, are "artificial wombs." Childbirth the old-fashioned way and the legal complications of surrogate mothering may become rare. (Cloning research may also lend to the "wild card" of life extension, as discussed by John L. Petersen in his essay "What Will Your Future Look Like?" in volume one of this series).

SUMMARY

Many of the old forms of the family will undoubtedly survive, as they have for hundreds of years, yet they will probably be joined soon by other quite novel forms—including communal and same-sex marriages, single-person households coexisting alongside large blended family households, and so on. As Joseph F. Coates writes in essay seven of this volume, "Change can be our friend; not always and not necessarily a threat to our future."

How soon, then, can we expect to see the fruition of some, if not all of my ten forecasts? I must be honest with you. Like many of your parents, I am a baby boomer. I thought many of these changes would happen in my own lifetime. But I underestimated social and cultural drag effects that slow, and even fight back change. So, I must counsel patience—and also activism: You would do well to push for the best of the possible changes I have highlighted—and promote improvements in the others all of your life.

QUESTIONS TO PONDER

1) What "wild-card" development might change your career goals to make you over into a housewife or househusband?
2) Walk an imaginary mile in the shoes of others. Can you imagine yourself in an untraditional occupation—for young women, as a long-haul trucker; for young men, as a maternity ward nurse; for young men and women, as someone preferring a same-sex person as a partner? What would your "pretend" lives be like?
3) Would you prefer to seek counseling from a face-to-face counselor or an on-line counselor? Why or why not?
4) Verne Wheelwright's "Your Map to the Future" essay in volume one of this series presents a set of stages of life to help the reader plot his or her "map to the future." How might women experience those stages differently than men?
5) How might I have underestimated the social and cultural drag effects that might impede change in your lifetimes?

NOTES

1 A body of law that seeks to uphold equality of rights without regard to sex.

2 "San Francisco Mayor Says 'I Do' to Same-Sex Marriages," http://us.cnn.com/2004/ALLPOLITICS/02/20/ip.pol.opinion.newsom/
3 http://en.wikipedia.org/wiki/Polygamy.
4 While the avoidance of jealousy and its consequences is often cited when *polygamous* relationships are outlawed, in a more practical vein, it has more to do with the avoidance of dealing with the property rights and codifying the legal responsibilities of multiple partners. Some Native American societies had practiced polygamy for countless centuries. Some religious sects, like the Church of Latter-day Saints (the Mormons), in their formative period, embraced polygamy as practiced by the patriarchs of the Old Testament. Pockets of polygamous Mormons can still be found in Utah and adjacent states today. The Mormons were not the only ones practicing polygamy during the early days of the United States. On the American frontier, it was relatively easy for a man who moved around with his work to have, in practice, a number of wives who never found out about each other. It has only been recently that this one-sided covert practice of polygamy has been harder to carry out (because of the ease with which we can find out information about one another).
5 A friend of mine in Alaska who had several wives in succession had a family reunion photo taken of all his wives, children, and in-laws from the different marriages. So large was his "family," that the picture resembled a regimental photograph from World War I.
6 Already the legally acceptable reasons for divorce in the United States have expanded to accommodate the demand of the public for the dissolution of their marriages. In many states, until recently, the range of rationale was limited to physical abuse, mental cruelty, desertion, and alcoholism. A reason like incompatibility of careers was considered frivolous. Divorce was time-consuming and expensive—and still is under certain conditions, where considerations for care of dependent children and property division are at stake. The market for *ease of divorce* probably can be attributed to rising mobility of Americans that combine the effects of the United States' transportation infra-

structure and changes in the economy that create the differential productivity of different regions of the country at different times. The legal system has been compliant in serving the divorce market because long-distance and commuter marriages bring their own complications.

7 All in all, I foresee a kind of Age of Discovery in relation to the possibilities of marriage and divorce. Alas, I am not as optimistic in the area of marriage and family counseling. As a sociologist who has held positions requiring me to perform various kinds of couples, marriage, and family counseling under other credentials that I have, I find the counseling profession languishing in a kind of Dark Ages. For example, many marriage and family therapists coming into counseling today don't know the differences between working-class and middle-class families that are needed to shape their therapeutics. On the verge of a Renaissance in human behavior application, the whole counseling and psychotherapy community took a giant step backwards in the 1960s when it began installing rigid licensing procedures that were meant to weed out quackery and incompetent clinicians. But, barred from establishing legitimate practices in counseling and psychotherapy were those trained as sociologists, as well as certain types of psychologists, including social psychologists. As a result, knowledge useful to counseling and psychotherapy has not made it across the licensure divide from those disciplines.

REFERENCES

Collins, Randall, and Scott Coltrane. *Sociology of Marriage and the Family: Gender, Love, and Property*. 4th ed. Chicago, Ill.: Nelson Hall Publishers, 1995.

Hauser, Philip M., and Otis Dudley Duncan, eds. *The Study of Population*. Chicago, Ill.: The University of Chicago Press, 1959.

McKie, Robin. "Men Redundant? Now We Don't Need Women Either": http://observer.guardian.co.uk/international/story/0,6903,648024,00.html (accessed 10 February 2002).

Nolan, Patrick, and Gerhard Lenski. *Human Societies: An Introduction to Macrosociology*. 9th ed. Boulder, Colo.: Paradigm Publishers, 2004.

Roszak, Theodore. *America the Wise: The Longevity Revolution and the True Wealth of Nations*. Boston, Mass.: Houghton Mifflin, 1998.

Part Two

ON FINDING MYSELF—TOMORROW

Unless you are going to do something beyond
what you have already mastered,
you will never grow up.
—Ronald E. Osborn

Ethnic identity is in flux, as perhaps it always is in nontraditional societies like ours, with their incentives in favor of assimilation and their contrary endorsement of pride in roots. Turmoil tomorrow is likely to be greater even than today, given the demographic trends well underway: For example, by 2050 Hispanics, our nation's largest ethnicity, are expected to increase their ranks by 188 percent, and become roughly 25 percent of all Americans, up from 13 percent at present.[1]

Essay one explores several implications of shifting ethnic numbers and well-being, the better to challenge the reader to take a position regarding promotion tomorrow of ethnic "mutuality and living wholeness." Another forecaster of a similar persuasion contends "the most intriguing question about, say, the year 2025 is not how America will change its Latin, Asian, and other ethnic sub-cultures, but how they will change America."[2]

The second essay should be understood against a backdrop of an America that by 2050 is likely to have racial minorities equal in number to 50 percent (up from only 31 percent at present); a very different America from this one (much more like the planet).[3] The essay salutes the arrival on the scene *already* of blacks who may prove forerunners of a radical change in racial identity, blacks who just might show the way to a finer America and world.—Editor

NOTES

1 C. Genaro Armas, "Census Sees Whites No Longer in Majority in U.S. after 2050," *Philadelphia Inquirer*, 18 March 2004, A-4.
2 Peter Schwartz, *Inevitable Surprises: Thinking Ahead in a Time of Turbulence* (New York: Penguin, 2003), 61.
3 Armas, *op. cit.*

■ Essay Four ■

THE DEMISE OF ETHNIC DIFFERENCES: CULTURAL IDENTITY AND THE INTERNET

Caleb Rosado, Ph.D.
Director, Urban Economic Development Graduate Program, Eastern University

The Internet, the most extraordinary engine of social change since the military conquest of nations, is helping us remake our ethnic and cultural identity. In place of a tired old focus on the color *of* people, we are experimenting with a healthy new focus on the "color" *in* people—on their remarkable systems of thinking and the resulting behaviors. This revolutionary change—still an optional one—has extraordinary significance for the rest of your life.

A BORDERLESS WORLD

Traditional approaches to human identity have focused on external issues of a biological, cultural, gender, or socioeconomic nature. We have rushed to label and prejudge people by their appearance, style of life, sex, or social class. Those of us who continue in the twenty-first century to categorize others on such a superficial basis may be blindsided by deep-reaching changes already looming on the horizon.

The Internet is altering the way we see ourselves and others, and the way our minds work. In ages past, where national and group boundaries were important for socioeconomic, religious, ethnic, and political reasons, identity with your own group, culture, and nation was vital. Passports—even in the nonpaper form of membership in this or that group—were essential for proving who you were, for entry, and for residence.

Today, the Internet has not just created fluidity across

traditional boundaries; in many situations it is actually obliterating them. I do not need my passport to cyber-travel to the United Kingdom and access the Bodleian Library at Oxford University. After writing the previous sentence I stopped and immediately traveled on-line to Oxford to access their main library to make sure that I had the correct spelling—I had missed the "i"—all within fifty seconds. No passport, no Concorde, no getting out of bed where I was doing my writing.

In a previous time period, I would not even have gotten through the door at Oxford due to color, class, culture, and country differences. Now all that is in the past. I went in and out in a matter of seconds and no one knew the difference, for none of those differences mattered. This is a small example of how the Internet is moving society toward a borderless world where all the old socially constructed boundaries are increasingly becoming irrelevant.[1]

The new variables that connect and divide people are their systems of values, their interests, the limits they set on their thinking and possibilities, and the new networks they forge with others who share the same interests and concerns in the cyber-world of the World Wide Web (WWW). On the Internet, one does not know with whom one is communicating (whether they are black or white, from Sweden or from Africa, a nine-year-old computer wizard or a Nobel laureate, is mono- or multilingual, a lesbian or a homophobe, an atheist or a Buddhist—the variables are endless).

What matters in the WWW are not the old visible boundaries. In the past, they preempted people from entering or expressing their ideas even before they could open their mouths—due to the mental scripts of the gatekeepers of power. The Internet trumps such twentieth-century superficial factors by creating a new sort of "affirmative-action" program. All, irrespective of who or what they are, have equal access as long as they can cross the Digital Divide and "get themselves connected." This is what really matters now—*connectedness*.

Here is where schools are already playing a key role. Education, K–12 and beyond, is now being "virtualized."[2] Students can and are taking on-line courses with students anywhere in the world. Virtually any course can be, and many will be, virtualized via the Internet. "Designer curriculums" are already in place. At leading high schools here and abroad, students design their own course of study. With wireless connection becoming widespread, learning will take place from anywhere in the world according to a student's interests, needs, and schedule, rather than the traditional 8 A.M.-to-3 P.M. classroom schedule.

CELEBRATING DIVERSITY OR LIVING WHOLENESS?

Two other factors associated with the Internet are also influencing new identities. The new "multiracial" category of the 2000 U.S. Census with which nearly seven million people identified themselves is also watering down ethnic/racial/cultural identities, eventually to the point where such differences will matter very little.[3] Latinos, for example, have been a multiracial people ever since Columbus' sailors landed and began the process of miscegenation with the indigenous populations. And there are very few blacks in the United States who can honestly claim to be one hundred percent African.

The Human Genome Project, with the data readily available on the Internet, is the other factor that is redefining who we are as human beings. The data show that all human beings have only some thirty thousand genes, and every person on Earth shares 99.9 percent of that same genetic code with all other people. Such findings demand that the focus must shift from superficial issues of race, color, and ethnic divides to the deeper levels of operational value systems within people.

"Celebrating differences" cannot be the ultimate goal or final stage of human relations endeavors. If such is their end, in the long run these efforts—as good as they appear to be—end up promoting racial/ethnic/gender divides that further alienate, fragment, and frustrate our social well-being. Focusing on what differentiates people at the expense of what they have in common does not further the cause and efforts for unity. To focus on the one-tenth of one percent of human differences, and not on these deep decision-making systems, is to focus on the gnat on an elephant's back and miss the whole elephant.

Successful human relations efforts must not only address factors that make people distinctive, but even more so, what they share that makes them similar. This is where the colorful spiral of emergent, unfolding behavior systems that shape human conduct and the way people see the world comes in. For from these differing value systems emerge, not only intolerance and discrimination among some, but also compassion and acceptance of human differences and the factors that create unity and human connectedness among others.

Unity is not synonymous with uniformity, neither is diversity synonymous with separation. The solution to the tension is to respect and value diversity while working for unity and wholeness, otherwise exclusion is the result. Thus, the strength of a nation or organization lies in *Living Wholeness*—the balancing of the two elements of unity-in-diversity through the valuing of wholes and parts.

What will make this possible is a moving away from a fragmentary worldview that treats the wholeness of the human family and social reality as inherently divided, fragmented, and disconnected; separating people into races, nations, cultures, classes, with each part regarded as "independent and self-existent" (David Bohm, *Wholeness and the Implicate Order*). The result is a fragmentary, atomistic, binary, mechanical worldview that divides what is indivisible—the oneness of the human family—and unites what is not capable of being united (groupings of people separate from the whole).

As the saying goes, "You cannot get there from here." The issues of racial, ethnic, and gender divides cannot be solved if one begins from a fragmentary view of the world. One needs to shift to one of "wholeness"—a holistic framework that acknowledges the interdependent yet autonomous nature of our social reality. The former comes from a Newtonian/Cartesian worldview: linear, binary, and mechanical. The latter comes out of nonlinear quantum physics, with its holographic view of the universe, where the whole is not only in the parts but bigger than the sum of the parts.

This Living Wholeness is a lifestyle unique to the twenty-first century; for never before in human history have so many different cultures and ethnic groups, with differing systems of thinking and "mindware," come together in the major urban centers of the world to form some semblance of social existence and convivialness.

THE INTERNET GENERATION

In terms of human social identity, a new generation with a new identity is emerging—*the Internet Generation*—one that did not directly experience the Civil Rights Movement and segregated living; one that operates at its own level of human existence, acceptance, and thinking, with a different set of operational values. Many members (though not all, of course) do not share the interest in social labels with others who may "look like them" and thus who may want to be identified by their Blackness, their Chicanismo, or their being a "model-minority."[4]

Where African Americans are concerned, for example, Nat Irvin, II, calls this new generation of blacks the "Thrivals." They are a "poststruggle generation of Africans and African Americans, characterized by a keen awareness of and appreciation for multiculturalism, globalization, and change. Drawing on strengths derived from historical victimization and oppression, they actively participate in global economics and politics and are a strong influence on social change. Individuals are typically competitive, critical, savvy, and educated, and have an outlook that is international and multicultural."[5] [Editor's Note: See the essay that immediately follows.]

Many members of this new Internet Generation—black, white, whatever—will be more impacted by its global cyber culture and individual achievements than by past bio-psycho-social-spiritual factors that forged the identity of their parents. Tiger Woods, with his "cablinasian" identity, is just one example.

NO MORE FRICTION?

The changes awaiting us in the near future will invite us to come to grips with the essence of our common core humanity, one that we tend to deny or hide behind cultural armor and image-enhancing facades of racial, ethnic, and gender identities. This does not mean, of course, the old paradigm of ethnic and cultural identity will soon disappear, for it won't. Ethnic wars will still rage and cultural differences will continue to dominate racial discourse, in great measure due to the values people uphold or are fighting off, and the self-interests they desire to protect and garner for themselves.[6]

For many, however, the core question begs—"Who am I?" as a human being; beyond being black, brown, white, Jewish, Arab, Irish, Buddhist, Christian, or whatever other sociocultural labels we may identify with. These labels reflect the objective "Me" (the way I and others perceive me), but not the subjective "I" (the essence of who I am as a human being beyond societal impositions). It is in the "I" where our similarities as humans are found, and where systems of change will increasingly focus.[7] As Howard Thurman said in *Creative Encounters*: "It is my belief that in the Presence of God there is neither male nor female, white nor black, Gentile nor Jew, Protestant nor Catholic, Hindu, Buddhist, nor Muslim, but a human spirit stripped to the literal substance of itself before God."

SUMMARY

Any understanding of diversity in this new millennium needs to consider how the new culture of the Internet can and probably is already shaping a new "cultural" identity that transcends old bio-cultural ones.[8] For as humans we have more in common with what unites us than with what divides us. In this Third Millennium, looming sociotechnological changes birthed by the Internet can aid our human minds to bring about a global transformational shift from competition and fighting for "rights" to mutuality and living wholeness—a shift that will salute connectedness. Care to help forge this finer Future?

NOTES

1 A new Web site has emerged, Words Without Borders (www.wordswithoutborders.org) that seeks to obliterate cultural borders that create "cultural blinders," by bringing literature and commentary from all over the world to the American public via the Internet.

2 Samuel L. Dunn, "The Virtualizing of Education," *The Futurist*, March–April 2000, 34–38.

3 These identities are socially constructed and are not part of our biological makeup. They all played a key and vital role in the nineteenth- and twentieth-century social movements from Civil Constraint focused on Slavery, to Civil War focused on Freedom, to Civil Rights focused on Equality. And yet the slavery-to-equality range of issues is as far as most conversations, committees, and commissions on race have gotten. The result has been a massive struggle for human rights and privileges, most of it to be lauded. In some case, however, the medicine has been worse than the disease. The backlash against multiculturalism and affirmative action, which has now moved to a campaign against ethnic studies programs in university campuses, has been in some cases an attempt to provide a corrective to extreme policies of political correctness. Yet, in other situations, it has been an all-out war of annihilation of effective programs.

4 In 1932, William F. Ogburn, the former president of the American Sociological Society, addressed such tensions brought about by social change by showing how the material aspects of culture (in terms of this discussion, the Internet) change faster than the nonmaterial aspects can keep up with (our value systems, the way our brains are presently wired). He called these periods "cultural lags," and the result can be severe internal conflict, both in the person and in the nation. The present national reality, oriented toward ethnic/cultural identity and flying in the face of the changes the Internet is bringing, is an example of such "cultural lag" and the psycho-social warfare it creates over culturally constructed differences. William F. Ogburn, *Social Change* (New York: Viking Press, 1932). For an understanding of where racism and ethnic conflict is headed into the future, see Caleb Rosado: "Memetics and the Multiple Futures of Racism: Beyond

the Myth of Race toward a New Paradigm for the Third Millennium." *Futures Research Quarterly*, Spring 1999, vol. 15, no. 1, 29–62.

5 Nat Irvin II, "The Arrival of the Thrivals," *The Futurist*, March–April 2004, 16–23.

6 The recent racial/ethnic riots throughout England and Europe make this clear that ethnic self-interests are not going to go away soon. Yet focusing on outward differences between groups, rather than on thinking systems, as much diversity and race relations training tends to do, will do nothing to ameliorate the situation. Thus, among some groups, depending on their level of operation in the spiral of human development, racial and ethnic identity will intensify due to worsening conditions. For others at other levels of existence, it will simply disappear as other factors take on greater importance. It was philosopher José Ortega y Gasset who reminded us that such attempts to hang on to "nationalism" in the face of inevitable change for inclusiveness represent the "last flare," the "last sigh," which are often "the longest" and "the deepest." And these tend to intensify "on the very eve of their disappearance." José Ortega y Gasset, *The Revolt of the* Masses (New York: W.W. Norton & Company, 1957).

7 Too many of us tend to be concerned with the "Me" and with all its self-enhancing entitlements with which we dress and prop up our otherwise fragile egos and dishonest behaviors. Even the virtual identities and e-consciousness the Internet is creating can become a new source of conflict if we forget that as humans we have more in common with what unites us than with what divides us. However, whether there will be an intensifying for some and inclusiveness for others depends on the deep-level value systems at which people operate, the forces of change they are addressing, and how open they are to theses changes.

8 The key factor here is what lies behind this new identity. It is not the old ideas of the color *of* people, but the color *in* people—i.e., the colorful memetic systems of thinking and the resulting behaviors these produce—which have always generated most of the conflict.

■ Essay Five ■

THE ARRIVAL OF THE THRIVALS! ARE YOU A MEMBER OF THIS NEW TRIBE?*

Nat Irvin, II, D.M.A.
President, Future Focus 2020, Babcock Graduate School of Management, Wake Forest University

A new and powerful force is beginning to emerge from within the black community and is in the process of changing forever the future of blacks in the United States and around the world. The world is witnessing the arrival of a new urban tribe that I call the *Thrivals*. They are the first generation of blacks who will aggressively compete in the battle to shape the images, ideas, and the future of global culture; business and commerce; science and technology; education; politics; the environment; the arts; and the role of nongovernmental organizations.

Thrivals are critical thinkers, technically adept, worldly, sophisticated, entrepreneurially driven, highly competitive; they are able to see the world through a global lens unfiltered by their own nationality, ethnicity, or culture. They believe international travel to be an essential part of one's basic human education. They are often bilingual—at the least. And when it comes to a test—any test—they are not about to complain. What they want is simply to ace the test.

For Thrivals, the world is theirs for the taking and they will not be denied. Instead of asking why, they ask why not? Quite simply, they view the world differently from their predecessors and many of their contemporaries, both white and black, and say quite emphatically, "I am not going to let you keep me from seeing what I see, just because you can't see it."

For Thrivals, there are no limits. And for the world's black community, they have arrived not a minute too soon. With problems such as far too many black males caught up in the U.S. criminal-justice system, the persistence of educational and earnings gaps between blacks and whites, and AIDS ravaging black communities worldwide, Thrivals are an answer to generations of prayers.

WHAT DRIVES THE THRIVALS?

Thrivals are the manifestation of a profound but subtle shift in awareness within the American black community and blacks worldwide: a shift from the consciousness of survival to that of thrival. For blacks, thrival represents a transformation of the soul, a transition from seeing oneself and one's community as being the victims of history and oppression.

Thrivals have moved from living in a survival mode, fighting for basic human rights, to embracing a new worldwide movement—a renaissance where succeeding generations, through imagination, self-determination, leadership, and legacy, see themselves as forces capable of shaping the future rather than being shaped by the forces of the future.

Among the forces driving the emergence of this new global tribe:

* Dramatic growth in the economic clout of American blacks.
* Continued increases in black educational achievement.
* Extraordinary escalation in the diversity of the United States, including within the black community itself.
* A decline in the significance of race and continued improvements in race relations in the United States.
* An overall decline of white majorities throughout Europe as a result of unprecedented changes in worldwide patterns of immigration between developing countries and developed metropolises.
* An explosion in the number of democratic nations around the world.
* The steady growth of urban areas worldwide resulting in a number of megacities.

* Growth and expansion of multinational corporations and affiliates. More than fifty thousand corporations with nearly half a million affiliates have emerged as a result of governments deregulating their economies, privatization of state-owned enterprises, liberalized financial markets, and trade.
* Increased international study among blacks worldwide.
* The global influence of the hip-hop culture phenomenon.

ECONOMIC CLOUT

Thirteen percent of Americans identify themselves as black or African American, a total of 34.7 million people, making blacks the second largest minority group in the United States, according to the U.S. Census Bureau. From 2015 on, more blacks will be added to the total population annually than non-Hispanic whites. By 2050, the black population is projected to reach 61 million, an increase of 83 percent.

As population numbers increase, so too will economic strength. Black buying power is expected to increase by some 34 percent from $688 billion in 2003 to $921 billion in 2008, according to the Selig Center for Economic Growth at the University of Georgia. In 1998, African American buying power alone was five times larger than the value of U.S. exports to the European Union. The Selig Center estimates that buying power for minorities (African Americans, Asians, and Native Americans) in 2008 will be more than triple the 1990 level, exceeding $1.5 trillion (a gain of $1.1 trillion or 231 percent).

The combined buying power of these groups will account for 14.3 percent of total U.S. buying power in 2008, and together will grow much faster than the white market. And as buying power strengthens, poverty continues to lessen its grip. The poverty rate for African Americans was 22.1 percent in 2000, according to the Census Bureau, the lowest measured since 1959.

But it is not just in buying power where economic growth is occurring. According to the Kauffman Foundation, black men between the ages of twenty-five and thirty-five with some graduate school education start more businesses than any other group in the country. In 2002, black women were owners of some 365,110 businesses, part of a larger trend, where in 2002 women of color started businesses at a faster rate than for other women and all businesses, according to the Center for Women's Business Research.

EDUCATIONAL ACHIEVEMENT

As Thrivals establish and strengthen their economic base and watch it grow by leaps and bounds, they continue to be aware of the importance of education and its place in the transition from mere survival to true *thrivalism.* In the United States and abroad, educational accomplishments are being made at all levels. At the same time, an awareness of the importance of an international outlook continues to expand.

The high school completion rate for African Americans increased from 68 percent to 76 percent over the past twenty years. Between 1980 and 2000, the percentage of African Americans between the ages of eighteen and twenty-four attending two- and four-year colleges rose to 37 percent for black men and 43.9 percent for black women. From 1990 to 2000, the share of blacks in the twenty-four-to-twenty-nine-year-old group with college degrees rose from 13.4 percent to 17.5 percent, according to the American Council on Education.

Thrivals believe in seeing the world as it is, not as it is told to them. As a result, education beyond the local level toward an expanded worldview is vitally important to achieving that goal. Over the past five years, the number of U.S. students who studied abroad for academic credit has increased by 55 percent, according to the U.S. Immigration and Naturalization Service. As a result of the sea of change in world demographics, barriers of race and ethnicity will not limit Thrivals. Indeed, Thrivals are the first generation of blacks of the future who will have a legitimate chance to compete for the best opportunities on the basis of their individual merit, skill, and preparedness. And they know it.

DESCRIBING THE THRIVALS

Think of the Thrivals as the group who sits out on the cusp of a tsunami, the leading edge of a global wave of multiculturalism. The essential cultural and economic base of the thrival phenomenon is in the United States, and its values are being transmitted and shared through four major venues: the electronic mass media, hip-hop culture, the Internet, and international study and travel.

Thrivals see themselves in the context of a global identity, not just in terms of a single nationality struggling between their American identity and their African diaspora roots. Instead, they use their respective bases of ethnic and cultural identity as a base of a pyramid, the top of which becomes something global in its cultural, and ethnic identity and understanding.

It would be a mistake to describe Thrivals as somehow being middle class, upper class, or elite, or even to use the distinction of economic class as a marker. Thrivals represent something much more. It is a global movement where blacks of all ages—worldwide—are primarily gathering in the world's large urban clusters near schools and colleges and centers of learning and commerce, in response to greater economic and social opportunities.

Manifestly international and urban in their geographical makeup, Thrivals could just as easily call cities like Kingston, São Paulo, Berlin, Houston, Kigali, London, and Rio de Janeiro home as they could Richmond, Virginia; Winston-Salem, North Carolina; or Camden, New Jersey. Their ideas of race, ethnicity, and culture are much broader and more nuanced than those of any generation in history, and they are much more likely to have friends and family from a wide range of ethnic and social groups, including Multiracials. They can be found throughout Europe, having emigrated from places like the British West Indies, Algeria, Senegal, Morocco,

Tunisia, Ghana, the Democratic Republic of the Congo, Eritrea, and Somalia.

They are part of what political scientist James Kurth describes as the "post-imperial immigration" pattern that has been under way for five decades, where the former colonial peoples migrate into the European metropolis. Tanzania, Ghana, Senegal, the Ivory Coast, Kenya, Ethiopia, and South Africa are among the countries that have seen the greatest increase in emigrants.

THRIVAL VALUES

Thrivals are loosely bound by a common set of ideas, cultural symbols, and heritage, including a single powerful meme: the struggle, having common ancestors who have labored to overcome oppression. Indeed, Thrivals are the first of future generations—the postmodern children—to emerge from the grips of the struggle, whether overcoming the legacy of slavery, Jim Crow laws, lynching, segregation, and legal disenfranchisement in the United States, or colonialism, imperialism, and corrupt government officials elsewhere. But rather than a hindrance, Thrivals see the struggle as a gift from the past, a kind of cold fusion of abounding energy that aids them in their drive to succeed in the future.

Not only do Thrivals share the struggle as a common historical bond, they also embrace a common set of universal principles they believe are the best ideas for economic and political development in the future. They share such Jeffersonian ideas as the belief in life, liberty, and the pursuit of happiness. They believe in capitalism, the free market, and the due process of law that supports commerce, the rights of women, universal access to public education, free speech, and freedom of the press. They see science and technology as the engine of social growth.

In other words, Thrivals are committed to the principles of an American style of liberal democracy, where the market, the rule of law, and a commitment to improving the lives of others regulate and influence commerce.

THRIVAL CULTURE

A thrival reading list may include Octavia Butler, Toni Morrison, George Orwell, Richard Wright, bell hooks, Guadeloupian novelist Gisele Pineau, Patrick Chamoiseau, or Caryl Phillips. Favorite titles range widely, including *Harry Potter*, *The Clash of Civilizations* by Samuel Huntington, *Middle Passage* by Charles Johnson, *The Long Walk to Freedom* by Nelson Mandela, *Erasure* by Percival Everett, *The Autobiography of Malcolm X*, and Danzy Senna's *Caucasia*.

Thrivals' musical tastes are equally global and wide-ranging, from reggae legend Bob Marley to New Age artist Enya to gospel musician Donnie McClurkin. And while Thrivals may celebrate the power of the hip-hop movement, they do not embrace hip-hop culture blindly. They have serious reservations about how the global mass media portray black women and how these often degrading images negatively impact the perceptions of blacks worldwide, both now and in the future.

Thrivals are fans of film directors who mix cultures and help open eyes to the world of different cultures. When asked about television, Thrivals commonly say they don't watch it, but when they do, they tune in to the Discovery Channel, History Channel, Animal Planet, CNN, BBC, and other news channels. On the Internet, they visit places like Africana.com, BlackVoices.com, BlackPlanet.com, and similar places for community information.

Thrivals are very careful about the people they select as their heroes. In addition to their parents, family members, and other close friends, their heroes include Gandhi, Nelson Mandela, Rosa Parks, Martin Luther King Jr., and the Dalai Lama. Thrivals believe that world leaders should exemplify the highest of ethical standards, regardless of who they are. Thrivals are quick to point out local heroes for inspiration and role models. Look for Thrivals to name their children in honor of a specific ancestor who was an early visionary or someone who worked hard making it possible for others to survive.

WHAT THRIVALS BRING TO THE FUTURE

The arrival of the Thrivals will pose a major challenge to the status quo within the black community as more members of this new global tribe begin to establish their identities in the social, political, economic, technological, and environmental arenas of the future.

For example, Thrivals may be ambivalent about such programs as affirmative action, which is seen by many American blacks as being essential to social and economic success in the future. Thrivals may choose to vote for or run as independent candidates, drawing from a base of support that is more appealing to a multiplicity of racial and ethnic groups, and in the process shun the Democratic Party, seen by many American blacks as the party more closely reflecting their values.

In education, Thrivals will stress greater preparation and the development of fundamental skills in math, reading, and writing. As they become parents, they will lead the global debate about black images the world over and will help spur a re-creation of a much healthier, prosperous, and a more values-centered black community. They will do this by appealing to successful examples of local heroes who, in the face of great odds, were able to overcome significant challenges to become leaders in their field. With a worldwide registry of contacts, they will be able to stand for public office and will be leaders in the development of some of the world's nongovernmental organizations, pushing for economic and social reform in the underdeveloped countries.

But not only will Thrivals influence the future of American public policy, they will also play a key role in shaping global politics, advocating the greater integration of developing-world nations into the global economy. Thrivals believe in social justice and equal rights and will be in a position to express support for such global initiatives as the wars on AIDS, hunger, and malnutrition, as well as for environmental regulations and new trade agreements with developing countries. Indeed, they will push for greater environmental safeguards for developing countries.

Because of their technical experience, they will seek greater use of biotechnology to help the countries of the world fight poverty and disease. They will actively support the United Nations and support the need for greater interdependence among nations. And by 2020, the fifth century of the American black experience, Thrivals will have had a significant influence on U.S. public policy, responsible in part for the election of the first black president of the United States of America—Sheniqua Nakada Herrera Kenyatta Hiawatha Jones.

Are you a Thrival?

* An earlier version of this essay appeared in *The Futurist*, March–April 2004, 16–23.

WEB SITES

"Global Trends 2015: A Dialogue about the Future with NonGovernmental Experts," produced by the CIA: http://www.cia.gov/cia/reports/globaltrends2015/

"People and Tomorrow's Markets: Serving Society." Source: "Tomorrow's Markets: Global Trends and Their Implications for Business" from the World Resources Institute, which is an independent nonprofit organization with a staff of more than one hundred scientists, economists, policy experts, business analysts, statistical analysts, mapmakers, and communicators working to protect Earth and improve people's lives: http://about.wri.org/

http://www.theglobalist.com/. Globalists: "Our goal is to explain the complexities of globalization to the people who ultimately drive economies everywhere. These are not the corporate CEOs and traditional policymakers. Rather, they are the citizens and consumers who live and work in the global marketplace—and want to make informed decisions. We provide timely, accurate analyses of key issues facing the global community in a straightforward, nonpartisan manner."

http://www.firstmatter.com/: Provocative thinkers!

American Demographics: http://www.demographics.com/

RAND Corporation: Overview. "For more than fifty years, decision makers in the public and private sectors have turned to the RAND Corporation for objective analysis and effective solutions that address the challenges facing the nation and the world. These challenges include such critical social and economic issues as education, poverty, crime, and the environment, as well as a range of national security issues.

Today, RAND researchers and analysts continue to be on the cutting edge of their fields, working with decision makers in both the public and private sectors to find solutions to today's difficult, sensitive, and important problems. Through its dedication to high-quality and objective research and analysis, and with sophisticated analytical tools developed over many years, RAND is engaged with its clients to create knowledge, insight, information, options, and solutions that will be both effective and enduring." http://www.rand.org/about/history/

The Milken Institute is an independent economic think tank: http://www.milkeninstitute.org/. "Our mission is to improve the lives and economic conditions of diverse populations in the United States and around the world by helping business and public policy leaders identify and implement innovative ideas for creating broad-based prosperity. We put research to work with the goal of revitalizing regions and finding new ways to generate capital for people with original ideas."

FURTHER READING

Books by Frank Fukuyama, Ray Kurtzweil, Jared Diamond, Sam Huntington, etc.

Also: *Long Walk to Freedom*, Nelson Mandela; *Before the Mayflower*, Lerone Bennett, Jr.; *W.E.B. Dubois, A Reader*, edited by David Levering Lewis; *The End of Blackness*, Debora Dickerson; *I Know Why the Caged Bird Sings*, Maya Angelou; and *The Parable of the Sower*, Octavia Butler.

Part Three

ON FINDING MY WAY—TOMORROW

Vision is the art of seeing things invisible.
—Jonathan Swift

As ours is the most religiously diverse of all nations, it is not surprising that religion looms large in the lives of Americans. Typical is 2003 research that found three in four college juniors thought religious or spiritual belief helped them develop their identity, nearly four in five prayed and discussed their religious beliefs with friends, even as 65 percent admitted to questioning their beliefs at least occasionally.[1]

Essay one looks over the horizon and shares several brow-raising forecasts. Changes are expected in intensity and character of religious identity, in the impact of fundamentalism, in the role of women, and from changing demographics, among other sources of turmoil. For example, there are already more members here of the Islamic faith than there are Episcopalians or Presbyterians.[2] And, as the United States is graying, it is worth noting that religion becomes more important the older an American gets.[3] As nearly seven in ten of us believe religion is losing influence in this country, the essayist helps us assess this "reading" and alternative interpretations.[4]

The second essay makes a case for updating the leading code of morality in the West, the Ten Commandments. Clearly, many pressing issues, such as the alleged Energy Crisis, are more of a moral, or theological challenge than only an economic or technical one.[5] (Related is the contention of a leading conservative journalist that "prayer should not be permitted in public schools, but maybe theology should be mandatory.")[6] If we are to measure up, we may want to go beyond the inherited wisdom of ancient clerics, and try our hand at writing thoroughgoing, modernized, and secular guides to conduct. Essay two offers an example, the better to help you take an informed position in the matter.—Editor

NOTES

1 Laura Randall, "Campus Trend: Getting Religion," *New York Times* (Education Life), 18 January 2004, 14.
2 Diana L. Eck, *A New Religious America: How a "Christian Country" Has Become the World's Most Religiously Diverse Nation* (San Francisco, Calif.: Harper, 2002). See also Pamela Pauls, "Religious Identity and Mobility," *American Demographics*, March 2003, 20.
3 Frank Newport, "A Look at Americans and Religion Today," Gallup Poll, 23 March 2004.
4 Barna Research, "Spiritual Progress Hard to Find in 2003," Barna Update (Internet), 22 December 2003. The data are from ten thousand adults interviewed by phone in 2003; the maximum sampling error is +/- 2 percent at the 95 percent confidence level.
5 See Vaclav Smil, *Energy at the Crossroads: Global Perspectives and Uncertainties* (Cambridge, Mass.: MIT Press, 2003).
6 David Brooks, "One Nation, Enriched by Biblical Wisdom," *New York Times*, 23 March 2004, A-23.

■ Essay Six ■

THE FUTURE OF RELIGION: THE DIVINE IN TIME

Ann Coombs
President, Coombs Consulting, Ltd.

A generation ago polite conversation did not include religion (too controversial, best left private). While this is not the case today, religion almost always stirs discussion and often intense emotion, even disagreement.

The controversy around Mel Gibson's 2004 film, *The Passion of the Christ*, taught us we might expect one another to have a better-than-ever understanding of spirituality in the near future. Certainly high schoolers, with their "mosaic" view of values, relationships, and spirituality—their openness and curiosity—are likely to have much to contribute. Eleven forecasts below may help get the conversation off to a good start.

GETTING BACK IN TOUCH

Looking out to 2010, it appears likely that an "ecumenism of spirituality" will probably gain strength. There is a growing openness to spiritual issues and signs of a related fatigue with secularization. Forgetting the extraordinary amount of religious material on the mass media (as in the Bible Belt states), we hear more and more spiritual discussions on mainstream talk radio, movies, and on family-hour TV series (as on popular TV family dramas like *Joan of Arcadia*, *Touched by an Angel*, and others).

More and more Americans seem to be on individualist spiritual quests. As recently as 2001, researcher George Barna found 84 percent of adults believed their religious faith was very important in their own life, and 70 percent claimed their own religious faith is consistently growing deeper.[1]

[Editor's Note: In 2004, a Gallup Poll found only 15 percent thought religion was not very important in their lives. Gallup Poll, 23 March 2004.]

RISE OF THE SECTS

New sects are a major reflection of, and gainer from, a steady turn in the United States to spiritualism. Protestantism's "new face," or its "post-denominational" or "new paradigm churches," would seem to some to be involved in nothing less than a second reformation. They are busy reinventing the way Christianity is experienced in the United States today.

More than one thousand new congregations are growing rapidly—pushing the envelope—and attracting large numbers of worshipers alienated from conventional religions. Parishioners boldly meet in nonreligious settings (including boating lakes, hotel meeting rooms, beachfront benches, poolside areas, etc.) and use Christian rock music, Eastern practices (meditation, etc.), an array of scents, dramatic color shows, and a variety of other nonconventional forms of worship to gain and hold interest. Nothing less than a transformation of religion may be taking place in the United States.[2]

Other various cults and religions abound in today's world. One example is Yawehism, a new religion that claims it can explain all of the existential questions of life. The fifty-one official new religious links that appear at www.meta-religion.com include The Last Testament Church, Friends Landing, and The New Christian Church of Full Endeavor, dedicated to the imperative of the naturally revolutionary progressive return of man to his inherent Universal Mind. Numerous counter-cults, occult, and esoteric organizations or movements are listed.

DECLINE OF FUNDAMENTALISM

Research suggests that while a large percentage of Americans claim adherence to a Christian faith, moral viewpoints consistent with conservative views are on a decline. Based on ten thousand interviews—of the 84 percent of adults who claimed to be Christian—there was an increase in 2003 in the proportions of Christian people who believe controversial items are "morally acceptable" behaviors: abortion (45 percent), adultery (42 percent), gambling (61 percent), pornography (38 percent), sexual relations between homosexuals (30 percent), and the use of profanity (36 percent).[3]

USING TECH TO TOUCH

In Western society in particular, a desire for spiritual connectedness has infused the Age of Technology. Information technologies offer "Interviews with God," churches on-line, and daily spiritual nourishment delivered to your cell phone. The growth of religion on the Internet has taken huge steps since the first "on-line ministry" in 1997. Now there are more than three thousand churches, cyber churches, virtual religions, and ministries on the Internet and in excess of twenty thousand electronic discussion groups. In addition, more than one hundred prayer circles exist on-line.

A desire, however, for embracing and sharing beliefs *in person*—a desire for connection of human spirituality—will probably always be strong. As young adults begin to reel from an "overload" of gee-whiz information and technologies, they may increasingly turn to religion for solace, balm, and guidance.

HEALTH AND SPIRITUALITY

Three-quarters of prayer groups now on the Internet are health related, and many practitioners in the medical field are being encouraged by patients and others to re-embrace God (long ago banished from certain hospitals and clinics in favor of sole reliance on secular use of pills, needles, and drugs).[4]

The importance of healing with mind *and* soul is once again gaining ground. More than half the medical schools in the United States offer "Spirituality in Medicine" courses to teach students how to talk with patients about faith and illness. Just a decade ago, only three schools offered such a course. This focus will probably gain strength in the future.[5]

DIVERSITY IN SPIRITUALITY

Buddhists, Sikhs, Jains, Zoroastrians, and other types outside the American Big Three (Protestant, Catholic, Jew) are arriving in ever-greater numbers, winning converts, and radically altering the religious landscape. (Buddhism, for example, has now become a mainstream American religion with about four million members, and it is growing. There are currently more than three hundred Buddhists Temples in Los Angeles, home to the greatest variety of Buddhists in the world.)[6] This new religious diversity is rapidly increasing and will undoubtedly continue to do so.

THE BIG THREE: ISLAM

As for the near future of the largest world religions, it would seem quite bright. Islam, for example, is the world's fastest growing religion with more than 1.2 billion followers.[7] Like all of the great world religions, it preaches compassion and justice. Indeed, most Muslims today do not regard jihad (holy war) as a central tenet of Islam.

Islam does not have an institutionalized, hierarchically organized clergy as does the Roman Catholic Church. Nevertheless, it has its own split between "clerics" and "modernists," the latter having since the 1920s been busy promoting modernizing reforms. Both blocs attempt to transform "traditionalist" practices of Islam, the first by purifying the religion of practices such as saint worship and faith healing that did not have Islamic roots; the second, the modernists, by attempting to construct a civil society outside of religious practice and authority.

The struggles between these two camps have created a rift within the Islamic world, particularly in countries like Egypt (where the reformers originated), but also in Jordan, Algeria, Morocco, Pakistan, Malaysia, and Indonesia. These states define themselves as "Islamic," but they have engaged in various forms of political, economic, and social "modernization." While they have at various times underwritten the rise of Islamist groups, they have also fought directly against them in their own version of our on-going "war on terror."

It is not difficult to forecast this struggle will wrack Islam for decades to come; far more difficult is forecasting how soon an accommodation can be achieved between the sides.

THE BIG THREE: CATHOLICISM

With more than a million followers, the Catholic Church is growing fastest in poorer developing countries. Over the years immediately ahead, the Catholic Church in North America will tend to lose both young and old people but will gain substantially from new immigrant joiners. Currently, the number of adherents in the Catholic Church in Europe is at a historic low, but it still remains, along with Islam, one of two culturally assertive religions with a global reach.[8]

Especially in the United States, the Church has been rocked by scandals involving cases of sexual abuse of minors, and charges have been pressed against more than four thousand priests.[9] It is expected that the scandals will dent but not bankrupt dioceses as lawsuits and lost donations take their toll on the Church. An image of looming financial collapse, however, ignores the fact that the Catholic Church in the United States is easily the richest in the world with untold billions of dollars in assets. Its 188 dioceses in the United States are distinct entities that cannot be sued as a group.[10]

As this is primarily a crisis in the North American church, it does not affect the majority of Catholic believers, and it will probably not result soon in major changes in doctrine.

Reformers—citing a crisis in a severe shortage of priests, particularly in South America—urge the Church to loosen up on celibacy and approve the ordination of women. The Vatican, however, does not want to alienate many millions of core supporters in developing countries to satisfy a relatively small number of liberal reformers in the West. So, securing radical reforms will probably take another fifty years, because opposition is so strong in the conservative Third World.

Reformers do not look to the next pope to radically shake things up, as they understand his role is that of a servant and not a master of doctrine. Since almost all of the 125 cardinals who will be voting for the pope's successor have been appointed by the conservative John Paul II, it is unlikely there will soon be any great departure in doctrine. So, in the near future Catholic priests will probably not be allowed to marry and/or nuns be accepted into a "priesthood."

THE BIG THREE: JUDAISM

It is impossible to think about the Jewish religion without considering Israel and the Israel-Palestine conflict. An end to that conflict will benefit both world Jewry and the world Arabic community. If the resolution is widely seen to respect Palestinian rights, fiery opposition to Israel (the all-too-ready fusion of anti-Zionism and anti-Semitism) will probably decrease, and with it hatred for Jews and Judaism.[11]

Anti-Semitism has been called "the longest hatred." Where this ties into religion is the belief that American Jews (and others elsewhere around the world) are seen as unquestioning supporters of Israeli policy (a total misrepresentation). Non-Jews who are sympathetic with the Palestinians believe this justifies their antagonism against all Jews.

Others who are eager to earn reconciliation here urge Israel, and Jews everywhere, to help form new and creative Jewish-Christian alliances as well as innovative alliances with peace-seeking Palestinians: "The world needs warriors of peace with wise words, patience and faith to outlast the guns of war and sands of time."[12]

POLITICAL PROBLEMS

What begins in religion commonly ends in politics, and the near future will undoubtedly see many more such controversies. China, for example, which before the mid-point of the twenty-first century may become the most powerful country in the world, continues to persecute on religious grounds. It has gone into Buddhist Tibet and banned photographs of the Tibetans' exiled spiritual leader, the Dalai Lama, and it strongly opposes Tibetan Buddhism.

Unofficial churches are sprouting up everywhere across

China. Escalation of religious persecution inside China has been documented by human rights groups and protests lodged by foreign governments, including President George W. Bush. The rise of Falun Gong (Fulun Dafa)—a spiritual movement—has faced China's harsh and continued campaign to decimate and discredit it according to ongoing reports of torture, rape, detentions, disappearances, and killings. More than one hundred thousand Falun Gong practitioners have reportedly been illegally arrested and detained, more than five hundred have been sentenced to prison terms of up to eighteen years, more than one thousand have been forced into mental hospitals, more than twenty thousand have been illegally sent to labor camps, and scores, if not many hundreds, have died in prison as a result of brutal treatment.[13]

In 1949, the Beijing government condemned Confucianism within its own borders and forced many of its followers into exile. During the Cultural Revolution (1966–1976) tens of thousand of monks and nuns were executed or sent to concentration camps, and six thousand monasteries and temples were destroyed.[14] China is now moving to repress "families" of religions and spiritual groups, and religious freedom as we know it in the West is clearly not on the horizon.

Elsewhere around the globe, and especially in certain countries in Africa and the Middle East, Christians are enduring persecution. One student of the subject believes "the mounting persecution of Christians eerily parallels the persecution of Jews ... during much of Europe's history."[15] Looking into the near future, it would seem likely that anti-Christian pogroms will soon provoke overdue debate and "shatter silences."[16]

Finally, we confront the unnerving case of a law passed in France in 2004 to prohibit Muslim women in school from wearing their distinctive head scarf (and other students from wearing similar signs of religious identity, such as yarmulkes, or large Christian crosses around one's neck, etc.). This attempt at enforcing secular blandness and uniformity seems to outraged opponents an exercise in antireligious arrogance, intolerance, and even racism (in the case of Islamic Muslims).

A grim forecast here is that many such clashes between overbearing secularists and besieged members of religious faiths may long blight the scene—and cry out for alleviation by young people committed to helping make a more "mosaic" future.

WOMEN—TO THE RESCUE

The role of women has not been great to date in *any* of the world's great religions, as they have been primarily male-dominated, but this is beginning to change—and looking out to 2010, much more progress can be expected. The depth of spirituality that naturally lies within women gives them the capacity to sufficiently hear one another without the strict rationality that often characterizes men's approach to many subjects, including religion.

While it may take decades, for example, before women can become Catholic priests, they increasingly play many key lay roles in Catholic affairs around the world. And there is momentum in favor of more gains by women in all religious faiths.

SUMMARY

Exciting and dramatic changes can be expected in the near future of religion. Where both individual spirituality and organized religions are concerned, young people—in particular—will have ever more opportunities to dialogue and work for peace and grace. As Melvin Konner explains, the common ground between all major ethical systems of different religions are "truth, respect for self, respect for others, and respect for something larger and more."[17] Let that be the common ground for high schoolers everywhere as they help influence the future of religion.

NOTES

1 The Barna Research Group, Ltd.: www.barna.org, "Spiritual Progress Hard to Find in 2003," (accessed 22 December 2003).
2 Donald E. Miller, *Reinventing American Protestantism—Christianity in the New Millennium* (Berkeley, Calif.: University of California Press, 1997).
3 The Barna Research Group, Ltd.: www.barna.org, "Spiritual Progress Hard to Find in 2003," (accessed 22 December 2003).
4 "Spirituality in Medicine: Ethics in Medicine," University of Washington School of Medicine: http://eduserv.hscer.washington.edu/bioethics/topics/spirit.html
5 Thomas R. McCormick, DMin, Faculty, Department of Medical History and Ethics, University of Washington School of Medicine.
6 Diana L. Eck, *A New Religious America* (San Francisco, Calif.: HarperCollins, 2001).
7 www.Adherents.com.
8 Ibid.
9 National Study by John Jay College of Criminal Justice.
10 Mark Clayton, *Christian Science Monitor*, 28 March 2002.
11 A Web site dedicated solely to balance what many believe is a consistent bias in the media against Israel: www.HonestReporting.com.
12 Phyllis Chesler, "Anti-Semitism: The Current Crisis and What We Must Do About It" as reviewed by Miriam Greenspan in *Tikkun*, November–December 2003.
13 Rabbi David Saperstein, Director, Religious Action Center of Reform Judaism at the Falun Gong Rally, Washington, D.C., July 2002.
14 "Persecution of Religion in Tibet," Position Paper, September 2001: http://www.savetibet.org/News/News.cfm?ID=143&c=27.
15 Michael Horowitz, Introduction, *Their Blood Cries Out*, by Paul Marshall, with Lela Gilbert (Nashville, Tenn.: W Publishing Group, 1997), XXI.
16 Ibid., XXIV.
17 Melvin Konner, *The Tangled Wing: Biological Constraints on the Human Spirit*, Revised edition (New York: W.H. Freeman & Co.; 2001), 481.

REFERENCES

Allen, Pamela Smith. *Awakening to the Spirit Within: Eight Paths.* Bloomington, Ind.: 1st Books, 2003. Examines eight different religions in the hopes of finding universal truths that promote peace and harmony rather than war and conflict.

Barna, George. *Real Teens.* Ventura, Calif.: Regal Books, 2001. Chapters are devoted to the religious beliefs and spiritual practices of teenagers. Among the discoveries reported in the text is that the top-ranked faith-related goal of teens is to have peace with God, which only ranked seventh in relation to all of the primary life and outcomes expressed by teens.

The Bible, Qur'an, Torah—These are the foundation and writings for all to read.

Chesler, Phyllis. *Anti-Semitism: The Current Crisis and What We Must Do About It.* As reviewed by Miriam Greenspan in *Tikkun,* November–December 2003.

Daily Wisdom (http://gci.gospelcom.net/dw/). A ministry of Gospel Communications, provides daily devotionals to cell phones and palm pilots.

Eck, Diana L. *A New Religious America.* San Francisco, Calif.: HarperCollins, 2001.

Marshall, Paul, with Lela Gilbert. *Their Blood Cries Out.* Nashville, Tenn.: W Publishing Group (Thomas Nelson), 1997.

Miller, Donald E. *Reinventing American Protestantism: Christianity in the New Millennium.* Berkeley, Calif.: University of California Press, 1997.

Peterson, Charles H. *Eclectic Pragmatism: Is There a God?* Bloomington, Ind.: 1st Books, 1998. A stirring and thought-provoking book on the relationship of science to God.

WEB SITES

www.adherents.com

Collection of more than 40,000 adherent statistics relating to 4,200 religions.

www.barna.org
Religious research group.
www.beliefnet.com
Popular interfaith site with discussions only for teens on religion and spirituality.
www.hollywoodjesus.com
A Web site devoted to spirituality in movies.
www.meta-religion.com
Multidisciplinary view of the religious, spiritual, and esoteric phenomena.
www.pbs.org/wnet/religionandethics
Current issues and resources—sites of the month are very informative.
www.pewforum.org
Seeks to promote a deeper understanding of the issues at the intersection of religion and public affairs.
www.pluralism.org
Committee on the Study of Religion—Harvard University.
www.religionsource.org
Find an expert on virtually any aspect of religion.
www.religionwriters.com
Postings every fifteen minutes on religious topics.

"You're kidding! You count S.A.T.s?"

■ Essay Seven ■

UPDATING THE TEN COMMANDMENTS: A DECALOGUE FOR THE TWENTY-FIRST CENTURY*

Joseph F. Coates

President, Joseph F. Coates Consulting Futurist, Inc.

Our world is more complex than could have been imagined by those who created our traditional religions. Accordingly, we wrestle with far more ethical and moral turbulence than is good for us.[1] To help us, every main religion, as well as many minor ones, has its equivalent of the West's Ten Commandments (also known as a *Decalogue*). The results, however, are either unconvincing, or do not relate to contemporary issues, or reflect tenuous interpretations often built up hundreds of years ago.[2]

The same limitations apply to counterparts of the Ten Commandments as to the Judeo-Christian tradition. For example, Muslims are not permitted to take interest on the use of money. And oriental religions advocate accommodation to one's present condition while implicitly denying the human capability to alter conditions and circumstances. Some faiths also in practice, if not in basic belief, make women secondary guests in the house of God.

Trying then to manage our world by force-fitting old ethical rules—such as the Ten Commandments—just does *not* work. They are narrowly personalistic and they emphasize prohibitions. They met a need of primitive societies, where change was slow, behavior was relatively uniform and stable, and innovation was rare. There was little need for positive thinking. So long as one adhered to customary practices, everyone survived and more or less got along well.

In our complex world, the situation is reversed. The problems we and our parents confront—issues relating to alcohol, driving, globalization, military service, rights to music, sexual activity, Walmart, and endless others—cannot be solved by negative rules. Old moral guidelines do not fit and cannot be made to do so. We must create *positive* frameworks in which to plan our future—to reduce the bad and stimulate the good.

UPDATING THE TEN COMMANDMENTS

A new Decalogue would certainly not replace the ten ancient mandates for us all. But it could guide us toward a finer understanding of the unprecedented issues unfolding around us. It could provide a route to better understanding *across* religious and cultural barriers and boundaries. And it could help meet the ethical-guidance needs of many who live outside traditional religious communities.[3]

Here are ten commandments for OUR times, ten that emphasize the *positive*—emphasize what we *can* do, rather than prohibit, constrict, and forbid us.

REALIZE THAT OUR ACTIONS OR FAILURES TO ACT DETERMINE OUR FUTURE

People alone are in charge of this world. No hypothetical or imagined powers, including devils and gods, can share responsibility with us. The failure to look ahead and the denial of our capabilities to shape or influence what happens is an invitation to disaster and a counsel of desperation.[4]

A core issue, for example, asks: *What do we collectively want to do with elephants now and in the future?*[5] Consider elephants. They have no standing, but I am sure that we can find a broad consensus that having elephants is far, far better than not having them, because they satisfy interest, curiosity, wishes, aspirations, desires, or provide entertainment for people. It is on that basis that we should intervene to preserve them or any other biological entity.

The potential utilitarian value that many argue as the basis for preserving a specific species (elephants) or a forest ecosystem has some force—but by no means an overwhelming one. One might decide that tropical forests or elephants are worth having even if they have *no* utilitarian or economic value. No one factor or nonhuman force (devils or gods) has any right or power to establish whether we do or do not act to preserve our common assets.

The implication here for youth is to *prepare to rule the world*, and for elders, *to prepare to have them do so*.

This first "Commandment," or injunction, is the ideological base for the nine that follow. Recognizing that we alone are in charge of our world can simplify and sharpen issues we and our successors must face in the decades ahead.

HONOR FUTURE GENERATIONS' RIGHTS, OBLIGATIONS, AND NEEDS

Encourage discussion, write essays, and promote media presentations of what future rights may be, such as rights to privacy, silence, leisure, and education. And work to maintain as many species and habitats as practical. Be sure there are not so many descendants as to exceed the ability of the world to meet their needs.

The need of people in the future implies for us the obligation to be sure that they can meet their needs. Put differently, our obligation to our descendents is to ensure that they have a broader scope of choices in life and lifestyle than we do. Planning our personal futures, which are, after all, short-lived, should not distract us from the longer-term effects on our personal and collective descendents.

RECOGNIZE OUR SOCIETAL AND GENETIC HISTORIES, AND WORK TO MEND THEIR FLAWS

There are failures or faults in both our social history and in our genetic origins.[6] Rather than reject them or glamorize them, we should acknowledge that change is our friend.

An example where our social history is concerned is the emancipation of slaves. Neither I, nor any of my ancestors, had anything directly to do with slavery. They arrived in the United States when slavery had already been abolished, and yet it makes sense for us today to try to fix whatever continuing disabilities or injustices came out of the long history of slavery and its social consequences.

Similarly, we each and all carry genetic defects. We may have to accept them for the moment, but let us determine how to eliminate, correct, or improve the situations they create. For example, many people in the deaf community now see being deaf as a special benefit. They organize and show resistance to research that could help change their condition. It is a destructive and divisive mistake to accept this genetically based defect as "normal" or "desirable." Such acceptance more or less assures no remedial progress will soon be made.

Characteristics we deplore in others and ourselves are all too often not based on free choices but determined by genetics or by the upbringing we have had. We must recognize what these negative features are and what their causes are before we can improve on them. Examples include pride, jealously, envy, lust, spitefulness, gossiping, grossness, anger, physical threats, and violence. They are in no way the works of a devil but misfortunes of our biological background, our education, our home life, or our social environment—and they are all increasingly correctable.

Change can be our friend; not always and not necessarily a threat to our future.

DO NOT DESTROY; ONLY IMPROVE

The good, the useful, and the beautiful are centerpieces of our world. Everything can have one, two, or all three of these characteristics, and we should work to make sure everything does.

Sometimes the features of utility and beauty are overlooked because they are either too familiar or even hidden from us. For example, one of the best representations of technological beauty and functionality is the automobile transmission. However, most of us see the beauty not in the innards of the automobile, but in the superficial fashions of the shell.[7] The more we recognize the good, the useful, and the beautiful, the more we will seek to make all things that way.

EXPAND OUR KNOWLEDGE AND DEVELOP OUR UNDERSTANDING OF ALL THINGS

Few if any problems, such as the future energy supply, water quality, youth pregnancies, or obesity, can be eliminated by a magic bullet. If we do not understand a complex issue, we are more likely to be captivated by the simpleminded remedy of the ideologue, the crowd-pleasing politician, or the earnestly strident special pleader.

The well-rounded person is less likely to act impetuously out of ignorance or prejudice, and to be more open to alternative outlooks, lifestyles, and choices for action.[8] By continually expanding our personal knowledge of all things, we do our part to foster a world of understanding and cooperation rather than one of hostility and violence.[9] Attention to all aspects of our nature, that is, being well-rounded, will give us happier, healthier, more interesting, and satisfying lives.

COVET ONLY WHAT WE HAVE EARNED AS A REWARD FOR OUR MENTAL OR PHYSICAL ABILITIES

In today's world, people can earn unimaginably large amounts of money, as the new multibillionaires have shown. But excess wealth breeds jealousy and greed. Therefore, we must seek and expect wealth only in proportion to what we have truly earned. Excessive wealth isolates the rich from the majority of humankind and their problems, needs, and desires. Unbounded wealth and its associated power is an intoxicant that may lead to a new barbarism, as it so frequently has in the past.

A future in which some few can gain unimaginably large amounts of money and claim their wealth as an entitlement raises issues of social justice. Also, it raises issues about the social value of passing that wealth on to successive generations or creating an institutional mechanism such as foundations where it may not be used for our best or most appropriate goals. Those institutions and those mechanisms may have been appropriate at another time of more modest scales of wealth but not necessarily today or tomorrow.[10]

HONOR ALL ROUTES TO THE TRUTH BUT NEVER SPECIOUS BELIEFS IN INFALLIBILITY

Truth comes from systemic exploration, not from abstract claims of direct knowledge or infallibility, nor from one's immediate and usually too-limited personal experience. Young people should be introduced to systems thinking in school, starting with human physiology, at the expense of mere facts and oversimplified accounts of complex things. Systems thinking should be the core of twenty-first-century education.

BE MODERATE IN ALL THINGS

This imperative—already widely acknowledged in terms of health and personal well-being—should be expanded to all our attitudes, behaviors, and actions. Moderation is the antidote to impetuosity, rashness, recklessness, and the destructive single-mindedness of assault, murder, and collective violence. By being more moderate in our behavior, we could inch more quickly toward a future peaceable kingdom.

RECOGNIZE OUR LIMITATIONS AS INDIVIDUALS AND AS SOCIETIES

Our limitations do not negate our rights or relieve us of responsibility. This implies an experimental approach to governance of our public and private affairs and frequent review of the effects of policies, laws, and regulations.

Imagine a futuristic plan to restructure K–12; or to redesign Social Security payments; or to build a new urban-to-suburban highway network; or to reduce addiction to gambling. Few of us believe public authorities alone could do a fully satisfactory job. But having all the stakeholders involved in setting goals and examining possibilities would invariably garner better results. Even better would be to recognize that our best *a priori* planning will have flaws, so we should design only flexible systems with which we can experiment, learn, and improve.

Democratic concepts must recognize that each of us has some capabilities *and* limitations in our knowledge and insight into what our actions should be. Personal, collective, and common actions should be based upon acknowledgement of our limitations. We should not act in haste if the price is to repent at our leisure.

TREAT INSTITUTIONS AND ORGANIZATIONS AS ENTITIES WITHOUT INTRINSIC RIGHTS

When we anthropomorphize institutions, such as schools, that is, when we impute natural rights and powers as if they were people—we isolate them from us and our needs. It makes them instruments for satisfying the private goals of their owners or managers. Schools that miseducate, and companies like Enron, WorldCom, and Arthur Andersen, demonstrate what harm mythic rights and privileges bring about—and will continue to do so until we re-establish the sovereignty of people over organizations.

Corporations as a legal creation are often used by people who own or manage them as a means of escaping responsibility for antisocial or highly questionable activities. For example, corporations act as if they were independent political entities, financing parties or politicians, whereas in fact, they are just reflecting the ideologies and interests of their top managers or principal owners. By anthropomorphizing business, especially the corporation, much mischief is done as those directly benefiting from business actions play up their firm's rights as *intrinsic.*[11] Organizational rights are a public *retractable* benefit given with reciprocal responsibilities to the people.

SUMMARY

Taken to heart, and pondered, this twenty-first century Decalogue can replace some inadequate ethical rules. It can complement the soundest of traditional beliefs and prove a creative source of new futures-shaping insights. It can be a valuable aid to shaping a finer tomorrow—to guide, manage, enjoy, and love a world that honors us all.

* An earlier and slightly different version of this essay appeared in *The Futurist*, May–June 2003, 68–69.

NOTES

1 Religious beliefs, the historic base of our ethical behavior, continue to erode in importance for at least two reasons: First, there are ever-fewer totally committed believers. And second, established formal ethical rules, primarily the Ten Commandments, emphasize two relationships we tend to devalue, that of man to God, and man to man. Instead, in our world, the newly dominant relationships are those of people to institutions, institutions to people, and institutions to institutions.

2 Members of religious communities are offered a wide spectrum of core beliefs. (Among Christians, for example, there are more than two hundred different sects).

3 Many nonbelievers have some carry-over beliefs to guide their behavior and to set expectations, but their activities usually are not solidly grounded in an ideological base. Their haphazard or noncoherent carry-over beliefs cannot help much in new unfolding ethical situations.

4 We surely do not want to regress to the charming pantheism of the Greeks, with every feature of the environment having its own god or goddess. This tendency shows up today in a watered-down form in which some people insist that trees have legal and moral standing. If so, what about lions, tigers, rats, or roaches? Any status they have is status we give them. To claim otherwise is infantile anthropomorphism, that is, seeing nonexistent human characteristics in the behavior of nonhumans.

5 A serious form of misleading spiritualizing often associated with the Gaia hypothesis sees the global system to be an organism, perhaps even sentient. The earth is an incredibly complex, but relatively robust and stable, system based on mechanisms that have evolved over billions of years, which respond to pressures on the system that tend to restore it to balance. There is no evidence of a need for a deeper, hidden explanatory force.

6 Leave DNA samples and full health records of yourself and family.

7 On the subject of automobiles, while driving we seldom pay attention to the routine beauty of a curving highway around a mountain or over hills.

8 A world of well-rounded people is likely to be a world of

understanding and cooperation. Many of our social problems such as crime, poverty, failing families, and degraded environments are associated with people who either never had the chance for a well-rounded existence, or whose lives are dominated by strong but narrow goals and indifference to the longer-term future as they concentrate on the short-term opportunities.

9 To broaden our understanding of the world, we must look, listen, touch, smell, feel, and taste more. We must learn about the environment, war, and peace, and we must encourage others to learn about similar things so that we can cooperate to build a better future.

Being well rounded means we will be less likely to operate in ignorance or prejudice and more likely to accept alternative outlooks and lifestyles or seek knowledge where we are ignorant. We should continually expand our personal and common knowledge and understanding of all things.

We are animals, therefore understanding of what our bodies need to remain healthy is essential to our well-being. Walk, dance, run, jump, play games. We are sentient, therefore look, listen, touch, smell, feel, taste, handle whatever we encounter to help our understanding. We think, learn, and make sound judgments. Therefore we must learn about what concerns us: food, clothing, shelter, work, the environment, governance, war and peace, and encourage others to learn about similar things so that we can manage our personal and collective affairs to create a better future.

10 Recall what Saint Bernard taught us: "Every rich man is a rogue or a rogue's heir." Billionaire rogues are a thousand times more dangerous than mere millionaire rogues.

11 Similarly, when corporations claim rights of privacy, these rights are not and cannot be intrinsic but only reflect the will of the people as revocably ensconced in law. It can be no violation of any intrinsic privacy rights to change the rules to acquire socially valuable information from them. The corporate claim, so abused in telemarketing and in advertising, to freedom of speech is a patent sham, because their messages are a consequence of decisions reflecting private views of their owners and managers.

REFERENCE

Coates, Joseph F., et. al. *2025: Scenarios of U.S. and Global Society Reshaped by Science and Technology*. Greensboro, N.C.: Oakhill Press, 1997.

Part Four

ON ENJOYING MYSELF—TOMORROW

Truly, Earth might be fair,
and almost all glad and wise
if we could only use our imaginations
to spin out scenarios of better ways to play.
—James A. Ogilvy,
Creating Better Futures:
Scenario Planning as a Tool for a Better Tomorrow

The essays in this section explore a wide range of fun-filled possibilities, from the proverbial *A* (as in artificial intelligence aids to video gaming) through *Z* (zeppelin flights and other exotic forms of travel). Along the way we learn much about Internet-based entertainment options, clever updates of sports equipment, biochip inserts into the bodies of athletes, suborbital junkets into space, tourist submarine tours of the sunken *Titanic*, and even parks whose animals lived millions of years ago but have been resurrected by DNA-based techniques.

Naturally, it will not be all fun and games, as much remains for remedial attention. In the near future, for example, advances in the biosciences may yet provide a pill or biochip to help those one in four Americans who told the Gallup Pollsters they sometimes drank too much.[1] Similarly, "couch potatoes" need help rapidly and permanently getting their weight back under control. Some 50 percent of adult Americans suffer from being dangerously overweight or obese, and since 1980 the proportion of overweight teenagers has tripled. As all are at increased risk of health problems ranging from some types of cancer to diabetes, various forms of active exercise would seem imperative in their tomorrow.[2]

All of us are likely to have to draw new boundaries, so great may be the temptation to lose ourselves in high-tech entertainment. Internet-based gambling, for example, centered in small Caribbean nations, is likely soon to greatly appeal, now that the European Union (EU) has ruled the United States cannot ban it.[3] While only 1 percent of Americans employed it in 2003, the percent may now rise dramatically.[4] One forecaster suggests that "when we go to work, the [virtual reality] game [we have been playing] will be installed on the inside of our glasses. We will gather at locations at specific times to meet others playing in our game. It's 'Matrix' to the max."[5]

Taken together, the five essays would seem too casual a read to paint only a dazzling fun-filled picture of tomorrow, but on a more careful read it will become apparent the essayists understand—and want the reader to understand—many trying choices and options are pivotal if we are to assure the game is worth the candle.—Editor

NOTES

1 As cited in Jeffery M. Jones, "Gambling a Common Activity for Americans," Gallup Poll, 24 March 2004.
2 Candy Sagon, "Obesity Is Epidemic, But Fat Isn't the Only Culprit," *Philadelphia Inquirer*, 20 January 2003, F-1.
3 Scott Miller and Christina Binkley, "U.S. Ban on Web Gambling Breaks Global Trade Pacts, Says WTO," *Wall Street Journal*, 25 March 2004, A2.
4 Jones, *op. cit.*
5 Faith Popcorn, Quoted in Suzanne Vranica, "Questions for …," *Wall Street Journal*, 7 January 2004, B-4A.

■ Essay Eight ■

ENTERTAINMENT IN THE NEW LEISURE ERA

Graham T.T. Molitor

President, Public Policy Forecasting/Vice President & Legal Counsel, World Future Society

All things considered, your generation is likely to gain more and more free time over your lifetime—certainly more than your folks ever enjoyed, and *much* more time away from work than your grandparents ever thought possible.

Around 2015, our country could become the first where more than 50 percent of a person's lifetime is available to pursue leisure activities. Choices will abound, and entertainment pursuits are likely to occupy much of our free time: lots of opportunity for a finer-than-ever life, provided you take care, stay creative, and take charge—especially of the entertainment component of your life.

Why so much time away from work? Because your generation is likely to choose more years of schooling and enter the workforce later than did your folks. Once on a payroll, you are likely to have shorter workweeks, more holidays, and longer vacations than ever before in this country.[1] Plus, you are likely to marry later than did your folks, and postpone having only one or two children (or none). You will probably enjoy more labor-saving appliances and devices of all kinds, and more personal services (from housekeepers to lawn keepers). And, thanks to this combination, you will probably welcome early retirement, and have an increased life expectancy within which to enjoy some truly Golden Years—perhaps on average as much as thirty or more of them.

What are the major entertainment choices ahead for your "free" time?[2] Consider the following eight forecasts carefully, always remembering that YOU have the power to chose, modify, or reject any and all of these possibilities: You, and *not* the entertainment sellers of this or that, should always be IN CHARGE![3]

SPECTATORSHIP

Cultural event activities occupy about one hour of our time weekly. Attending sporting events, not surprisingly, accounts for 40 percent of activities in this category—and most of it at the high school level. Movie attendance ranks second, taking up another 33 percent of the time segment. Then, another 15 percent involves fairs and amusement parks. The remaining 12 percent we put into attending concerts, museums, and other cultural events.[4]

You probably have been turned away at some recent entertainment attraction because it was sold out. Attendance at all kinds of events is up. It figures. Population is increasing and is expected to at least double over the next twenty-five years. Increasing affluence also increases financial ability to attend costly events.

So, my first forecast: You can expect more trouble getting to see just who you want, when you want, and from the seats you prefer. Sporting events aren't the only forms of entertainment that will increasingly sell out. Concerts (featuring rock stars to classical musicians), first-run movie theaters, major museum exhibits, and even White House tours are also bumping up against ceiling limits.[5]

Which is not to say that promoters are not trying. More than $16 billion was spent during the 1990s for construction of giant sports complexes for professional and college sports activities. This level of outlays represents a giant leap above and beyond the $3 billion grand total spent on such facilities during the 1970s and 1980s combined. Supply of seats, however, will not ever keep up with demand—not, at least, over the next ten or so years.[6]

One person's favorite may be another's dread. Ever hear of "shopper-tainment?" Yes, shopping malls and gallerias are a growing tourist/entertainment attraction. Shop until you drop, alas, has its devotees. The super-huge seventy-six-acre Mall of America, in Bloomington, Minnesota, although catering to inveterate shoppers, offers something for everyone. This gigantic shopper's paradise includes a twenty-three-ride amusement park (including a roller coaster), miniature golf course, aquarium, fourteen-screen theater, nightclubs, bars and restaurants, and nine family entertainment attractions (Camp Snoopy and others). More such remarkable temples of consumership are on the drawing boards and going into construction.[7]

TV'S SECOND WIND

The number of TV sets in the United States probably already exceeds the number of inhabitants. In 2000, there were 2.4 TV sets per household (and 5.6 radios). TV sets per family, boosted by mobile viewing screens, are expected to double by 2020. Vivid home theater giant TV screens unleash another new "must have." This new "electronic hearth" focal point is likely to supplant the family room and fireplace.[8]

Rising affluence and lower prices are increasingly bringing luxuries formerly enjoyed by a few within the reach of many more. Family rooms and entertainment centers are giving way to elegant, if not sumptuous, home theaters. Entertaining at home will probably increasingly be centered around home theaters, complete with luxurious loge seating, full wall screens, surround sound, popcorn machines, hotdog wagons, full-service bars, and more.

More elaborate home entertainment centers will get a boost as telecommuting grows. Hi-tech home office equipment will augment home entertainment gear. Combination at-home offices and entertainment centers will probably become the centerpiece of tomorrow's housing. Workers tend to favor splitting time spent on the job fifty-fifty between home and office. Growing numbers of workers already enjoy similar arrangements. This is a trend you can bet on!

More TV screens are cropping up everywhere. The current rage involves on-board TV sets for automobiles. These systems entertain youngsters and help soothe parents' nerves, especially on long trips. It makes sense to immerse motor-car environments in relaxing entertainment. It helps coping with bumper-to-bumper gridlock and soothes the ride. The way things are headed, everywhere you go, video screens will be omnipresent. Viewing time is destined to increase.

Far and away, whiling away "free time" before TV—fifteen hours weekly—is the Godzilla of all pastime entertainment. Time diary 1985 schedules revealed 123 minutes daily devoted to primary viewing. Secondary use occupied another forty-seven minutes. In addition, another eighteen minutes involved tertiary utilization. This 188 minutes amounts to nearly one-half of available leisure time. It will continue growing. America is becoming a "visual" culture and no other free-time activity comes close.[9]

Delivery of programming, however, is headed toward *pay viewing*, whether by satellite, cable, or other carrier. Rapidly rising monthly cable TV charges are beginning to turn away subscribers. As glitz and quality wane, free television may become scarcer and offer fewer choices (as the saying goes, "One thousand channels and nothing's on.")

PERSONAL COMPUTER USE SOARS!

Home computers are becoming a strong competitor to TV. The percentage of computers in U.S. households rose from about 10 percent during the 1980s, to about 54 percent around 2001. Personal communications devices accessing TV, radio, the Internet, and a host of other options provide "tailor-fitted" entertainment; the kind of programming younger people want. Home gateways to the information highway—Internet links to service provider networks—are projected to surpass $5 billion this year (2004). This level represents a huge jump from $1 billion grossed in 2001. Things are changing—fast.[10]

On-line entertainment takes many forms and caters to many diverse interests. Seeking hobby information topped the list in 2002. Second on the list, placing only a trifle lower, was browsing for fun. Next came downloading games, videos, and pictures. This was followed by playing games. Further down (in rank order) came listening to music/radio; visiting adult Web sites; downloading movies; and gambling. Time diary surveys have been conducted every ten years over the past four decades. Tabulations covering "home communications" reported an average of 4.4 hours during 1985. That number will continue to change and increase.[11]

The Internet, as it improves and grows, displaces many traditional activities.[12] Long-distance telephone companies, for example, report that e-mail, instant messaging, and Internet calling has displaced 30 percent of their business. Little wonder. It is far cheaper, faster, and more convenient and this trend will only accelerate.

WIRELESS PHONES

More than 60 percent of 15- to 50-year-old Americans are likely to carry wireless communication devices by 2007. By 2005, 40 to 60 percent of high-tech communications will be wireless. In 1990, there were only 5.3 million cellular phone subscribers. As of 2003, 147 million cell phones in use had nearly surpassed the 187 million wireline units. By 2005, the substitution of the newer technology will begin rapid ascent to dominance. Cell phones are projected to surpass wireline phones long before 2010.[13]

Right now, impromptu messaging is enjoying a heyday. Impromptu socializing and downloading diversions provide entertainment. Wireless phones are likely to account for more than 50 percent of logged calling minutes by 2006. What is making these mobile devices so desirable is the capability of merging most of the coveted communication equipment into manageable single packages. From an entertainment standpoint, growing capabilities as a single purpose multimedia device make wireless cell phones a "must-have" item. Single multimedia devices that slip into a pocket are what the future holds.

COMPUTER/VIDEO GAMES

Interactive entertainment is surging. As games become more realistic and affordable, play time will increase. A sign of the times is that computer/video game revenues have already surpassed that of movies. Business providers will follow the money. For example: *GoldenEye*, the James Bond thriller, cost $60 million to develop and grossed $106 million; Nintendo's version of *GoldenEye* cost a mere $4 million to develop and grossed $230 million.[14]

Eventually, old-fashioned board games, cards, and dice will fade and almost totally be supplanted by electronic versions. In similar manner, virtual reality versions of messy and costly paintball are certain to marginalize participating in the real thing. Interactive software programs teaching flying and driving armored tanks provide both entertainment and training. Military training in this manner is far cheaper than doing the real thing. Next, virtual reality will take over—big!

ON-LINE MUSIC

Sales of on-line music will grow by leaps and bounds in the years ahead. Low-cost recording and mixing gear has spawned at least thousands of home recording studios. Hi-tech electronics open up new opportunities for no-name musicians to do their thing and Internet publication facilitates distribution. In like manner, imminent advent of accurate voice-entry opens up opportunities for everybody to become a creative author, poet, playwright, or literary figure.

MOVIEGOING OR GETTING

In this entertainment-rich Leisure Era, you might think that theatergoing is waning. Admissions actually are increasing. Paid admissions rose from 1.02 billion (1980) to 1.64 billion (2002). Chances are good that the movie you want to see is available nearby. Multiscreen theaters that used to feature four to five screens, now boast eighteen to twenty.[15]

There are points of diminishing returns. Escalating prices may make a night at the movies too costly.[16] Many will prefer to view a film at home—on DVD, video, cable, or whenever network broadcasters run it. DVD will shift over to video-on-demand. With no special trips to acquire or return rentals, convenient and quick on-line digital deliveries will put an end to most video/CD rental shops.

On-line movies, music videos, and streaming video of all types grossed $65 million in 2001, and may exceed $2 billion by 2005, while becoming the overwhelming favorite mode to watch movies within ten to fifteen years. Drawbacks to overcome include time-consuming downloading and insufficient data storage. Exotic carbon nanotubes are likely to become the storage medium of choice. Memory chips storing terabits per square centimeter could be on the market by 2005.[17]

It is hard to imagine, but this storage density is one million-fold that of contemporary modules. Up to 40 percent of U.S. households (37 million) are projected to be using video-on-demand by 2007. A *New York Times* article reported that at-home black boxes costing about $100 will be capable of storing as many videos as a stocked video store. Times are changing.

One gains a sense of the momentous and growing dominance of the Leisure Era entertainment draw by considering top tourist destinations. The fact that 35.7 million Americans visited Disney attractions in Orlando, Florida, during 1998, while only 21.1 million visitors (1996) visited the nation's capital may be an indication of where the nation is headed.[18]

LIFELONG LEARNING

Thanks in part to our ability to enjoy better-than-ever distance-learning course offerings, we are likely to make more of living-room education, as in noncredit courses taken in our homes—courses custom-tailored to our learning styles. According to *Business Week*, by the end of 2004, one hundred million Americans are projected to be taking continuing education courses. Lifelong learners seek mastery of new skills, ranging from gourmet cooking to kung fu. And for some, these pursuits and interests can turn into lifelong careers: professional athletes, chess champions, animal trainers/vets, and yo-yo and Frisbee aces who become exhibitionists and paid demonstrators.

SUMMARY

You are likely to have 101 exciting new ways to enjoy more free time than any generation of Americans has ever known—and you cannot begin soon enough to explore and make rewarding choices among the options: Good entertainment beckons, and a healthy and happy life includes making the most of it!

NOTES

1 Graham T.T. Molitor, "Emerging Economic Sectors in the Third Millennium: Leisure Time Era Begins to Dominate U.S. Economy by 2015," *Foresight*, vol. 02, no. 04, August 2000, 370–371. Reduced workweeks that declined from 72 hours in the late 1700s to 34.5 hours, may decline to 30 hours by 2020. More vacation days; now averaging 12 in the United States, but 42 in Italy, indicates more are likely to come. Additional Federal holidays; now 12 in the United States, but 18 in other nations, portend increases. Leaves of absence (military duty, jury service, parental leave, medical purposes, funeral attendance, personal leave, sabbaticals, rest time, lunch time, coffee breaks, and so on) add additional time off. Earlier retirement, which dropped from 67 years (1950s) to a mode of 60 years (1994), is likely to decline to 55 years, a level already reached by males in The Netherlands.

2 _____. "The Leisure Era," *Executive Excellence*, November 2000, 11–12. As it involves amusement and diversion, an arbitrary short list includes: alcohol; boats and motors; entertainment; facilities; food and drink; gambling; hobbies and handicrafts; electronic home entertainment; event admissions; exercise equipment and services; information publications; lodgings; personal aircraft; pets and pet care; photographic equipment and supplies; power tools and lawn care; recreation activities, services and equipment; spas and retreats; sporting goods and equipment; sports clothing and footwear; travel and tourism; toys and games; vacation homes; vehicles (bicycles, scooters, snowmobiles,

motorcycles, camping trailers, etc.), and miscellaneous. Whatever the elaboration may be, it is vast.

3 _____. "Five Economic Activities Likely to Dominate the New Millennium," *Technology Forecasting and Social Change*, vol. 65, no. 2, October 2000, 244; "Here Comes 2015: The Onset of the Leisure Era," *Vital Speeches of the Day*, vol. LXVI, no. 20, 1 August 2000, 624–625. Entertainment is big business—U.S. government tabulations for 2002 show an increase to $620.1 billion. These totals do *not* include tourism which accounts for well over another $500 billion. Food services, accounting for 50 percent of all food spending, also is overlooked. Total gambling and wagering receipts in 1999 have been estimated as high as $650 to 700 billion for legal activities (plus another $100 billion in illegal wagering). Legal gambling receipts are projected to reach $1.5 trillion in 2005 and $3 trillion by 2200. On-line gambling, almost certain to be legal within the next ten years, will provide further increases. Entertainment sometimes lurks in the background throughout one's daily routine. Taking a shower or waking up in the morning may be accompanied by radio, TV, or a favorite CD. Broadcast media background music or a talk show host may accompany commutes. More than 90 percent of time spent listening to radio or recordings is a secondary activity while doing other things. Americans cram more tasks into given time slots by performing multiple tasks simultaneously. In-flight movies while away travel time en route to a ski trip. All the time, you might peruse assigned reading, write to a friend on your laptop PC, telephone your favorite uncle from a seatback-mounted telephone, sip a cola, converse with seatmates, and flirt with flight attendants.

4 John P. Robinson and Geoffrey Godbey, *Time for Life: The Surprising Ways Americans Use Their Time* (University Park, Pa.: Penn State University Press, 1997), 174.

5 Pollstar, "Most Expensive Concerts, 2001," covering January 1 to June 30, 2001. More than $508 million was grossed by just the top fifty concert tours during the first six months of 2001. Surprisingly, the highest ticket price—$119.76—was to hear the opera singer, Luciano Pavarotti. Tickets to see and hear Sting cost $53.49. The average ticket price for performers in the top fifty was $46.69, considerably up from $21.22 in 1991.

6 Warren Cohen, "Oysters, Scotch and Hoops," *U.S. News & World Report*, 15 November 1999, 92–93.
7 Shay Sayre and Cynthia King, *Entertainment & Society: Audiences, Trends, and Impacts* (Thousand Oaks, Calif.: SAGE Publications, 2003), 341–342.
8 Annie Groer, "Huge Screens, Reclining Seats and Cup Holders Now Showing at Home," *Washington Post*, 29 January 2004, H1, H5. Average cost ranges between $15,000 and 25,000. Following trends toward ultra-costly trophy kitchens and luxurious spa-sized master bathrooms, all the bells and whistles in home theaters can cost $100,000 to $200,000. No frills left undone versions run well over one million dollars. La-Z-Boy's overstuffed recliners replace "family room" couches. Complete with a built-in beverage cooler (twelve-can capacity), massager, heating system, phone and remote control, you hardly need to stir from a reclining repose.
9 Geoffrey Godbey, *Leisure in Your Life: An Exploration* (State College, Pa.: Venture Publishing, Inc., 5th ed., 1999), 73.
10 Martin Crutsinger, "Computers in Half of U.S. Homes," *Washington Post*, Fall 2000, 35.
11 Robinson and Godbey, *Op. cit.*, 125. Stretching definitions, each of these activities can be considered "entertainment"—some more so than others. Not many would consider the 2.8 hours weekly spent reading as entertainment. Most would consider the 0.4 hours weekly listening to radio and recordings as entertainment. Growing popularity of digital downloaded carry-around tunes will increase listening time. Almost everyone would agree that the 2.7 hours devoted to hobbies and the 2.2 hours spent pursuing sports and outdoor activities fall within the definition of entertainment.
12 Boston Consulting Group and NPD Group, "North American Internet Users Survey," *Time*, 17 April 2000, front matter. Respondents to a recent survey stated that internet communication and entertainment capabilities replaced a host of alternative activities. Letter writing was changed or reduced for 64 percent of those surveyed; long distance calling by 46 percent; watching TV by 43 percent (59 percent in another poll); reading newspapers by 24 percent (34 percent in another poll); magazines by 21 percent; and books by 19 percent. Other findings found that 13

percent of respondents wound up spending less time with family and friends; and 8 percent attended fewer social events. These impacts and incursions will continue to grow as more persons become comfortable with the new medium.

13 "Industry Watch: Entertainment," *IQ*, May–June 1992, 14. See also www.Kurzweilai.net.

14 Sayre and King, *Op. cit.*, 275. Making movies is expensive. Overall production-distribution costs of feature films nearly tripled between 1987 and 1997. What had cost $28.4 billion rose to $75.6 million, and the costs keep escalating.

15 *Statistical Abstract of the U.S.*, 2003, Washington, D.C.: U.S. Government Printing Office, Table No. 1225.

16 Ibid., Table No. 1225. Ticket prices that averaged $2.69 in 1980 and $4.23 in 1990, cost $5.39 in 2000. Flush customers are being enticed by "full-service" treatment. For $12 to $15 moviegoers get free valet parking, elevator service to a private floor, free popcorn, reservations in on-site restaurants, on-site baby sitting, and gourmet snacks and alcoholic drinks to take into private lounge seating areas. Little wonder it costs more and more to go to the movies.

17 *Business Week*, 27 October 2003, 72.

18 "How's Business Newsletter," Tallahassee, Fla.: FLAUSA Visit Florida, 2000. See also *The World Almanac and Book of Facts*, 2001, 714; Evans, *Washington Post*, 1998, E1, E3.

REFERENCES

Aron, Cindy. *Working at Play: A History of Vacations in the United States*. New York: Oxford University Press, 1999.

Boston Consulting Group and NPD Group, "North American Internet Users Survey," *Time*, 17 April 2000, front matter.

Cohen, Warren. "Oysters, Scotch and Hoops."*U.S. News & World Report*, 1999, 92–93.

Crutsinger, Martin. "Computers in Half of U.S. Homes." *Washington Post*, Fall 2000.

Dumazeider, Joffre. *Toward a Society of Leisure*. New York: The Free Press, 1967.

Evans, Judith. "Revolt Results in Changes in White House Tour." *Washington Post*, 1998, E1, E13.
Godbey, Geoffrey. *Leisure in Your Life: An Exploration*. 5th ed. State College, Pa.: Venture Publishing, Inc., 1999, 216–218.
Groer, Annie. "Huge Screens, Reclining Seats and Cup Holders Now Showing at Home." *Washington Post*, 29 January 2004, H1, H5.
"Industry Watch: Entertainment," *IQ*, May–June 1992. See also Kurzweilai.net.
Larrabee, Eric, and Rolf Meyersohn. *Mass Leisure*. 3rd ed. Glencoe, Ill.: The Free Press, 1960.
Molitor, Graham T.T. "Here Comes 2015: The Onset of the Leisure Era." *Vital Speeches of the Day*. Vol. LXVI, No. 20, 1 August 2000, 624–625.
_____. "Emerging Economic Sectors in the Third Millennium: Leisure Time Era Begins to Dominate U.S. Economy by 2015." *Foresight*, vol. 02, no. 04, August 2000, 370–371.
_____. "Five Economic Activities Likely to Dominate the New Millennium." *Technology Forecasting and Social Change*, vol. 65, no. 2, October 2000, 244.
_____. "The Leisure Era." *Executive Excellence*. November 2000, 11–12.
Robinson, John P., and Geoffrey Godbey. *Time for Life: The Surprising Ways Americans Use Their Time*. University Park, Pa.: Penn State University Press, 1997.
"Rock the Register." *U.S. News & World Report*, 23 July 2001. Adapted from Pollstar, "Most Expensive Concerts, 2001," covering January 1 to June 30, 2001.
Sayre, Shay, and Cynthia King. *Entertainment & Society: Audiences, Trends, and Impacts*. Thousand Oaks, Calif.: SAGE Publications, 2003.
Streisand, Betsy. "A Large Popcorn and a Martini, Please." *U.S. News & World Report*, 21 August 2000.
The TBR Index of Leading Economic Indicators. New York: Travel Business Roundtable, 12 April 2000.
Travel and Tourism: A White Paper. New York: Travel Business Roundtable, 2000.
The 2000–2001 High School Athletics Participation Survey. Indianapolis, Indiana: National Federation of State High School Associations, 2002.

The 2003 Statistical Abstract of the U.S. Washington, D.C.: U.S. Government Printing Office, Table No. 1225.

The World Almanac and Book of Facts. Mahwah, N.J.: World Almanac Books, 2001.

"I don't know, honey. What do you feel like doing?"

■ Essay Nine ■

PURSUING RECREATION IN THE NEW LEISURE ERA

Graham T.T. Molitor
President, Public Policy Forecasting/Vice President & Legal Counsel, World Future Society

How might recreation differ in the years immediately ahead—years likely to have more and different options than ever before? The sooner you begin to sort out your cutting-edge preferences, the more likely you are to make the most of them—including choices among scaling mammoth mountains, sailing azure blue waters dotted with emerald isles, scuba diving in undersea worlds, piloting airboats skimming swamps, or—laying about (with a good book) on a sunny sandy beach.

PHYSICAL ACTIVITIES AND HEALTH

High school- and college-aged persons—ages sixteen to twenty-four—indulge in more vigorous physical outdoor activities than older age groups. As Americans increasingly discover their flab, local workout facilities are enjoying a heyday. Regular exercise is estimated to add about one year to a person's lifetime. Health risks associated with inactivity help encourage exercise activities. Fitness activities involve at least twenty different kinds of activities and businesses catering to this contemporary interest are flourishing. So, it is easy to forecast ever more participation in formal exercise—of one kind or another.[1]

A boost here comes from more construction of giant $100 million year-around, man-made surf and snow indoor sports environments.[2] These facilities include indoor snowboard and surf entertainment/sports options. Winter sports can host snowboard halfpipes, and an ice skating rink, along with artificial snow "peaks," and slippery downhill runs. Man-made beaches, complete with surf, sand, and tropical breezes, allow swimmers and surfers to choose among wave pools of varying difficulty, regardless of the weather outside or the time of day or night.

SPORTS EQUIPMENT

Certain equipment changes are easy to forecast, e.g., old-fashioned Louisville Slugger wooden baseball bats have already given way to high-tech fiber and exotic gas-filled construction. PST (powdered-shelled technology) super-swats—aluminum, and carbon fiber and nitrogen-filled versions costing up to $500 are top picks.[3] Skiers benefit from all kinds of newly configured boards with exotic materials construction, including electronic torque adjusting members that enhance "carving" and control. Sleeker and slipstream clothing for swimmers, bikers, and racers promise to help enhance performance. Athletic shoes with wondrous enhancements for every fleet-footed move imaginable—from ballroom dancing to soccer (perhaps the most sophisticated hi-tech design imaginable)—will probably overwhelm markets with thousands of different configurations and choices.[4]

Tennis decision-makers, for example, knowing the sport faces declining popularity, assiduously readjust the game to suit older aficionados and to encourage less-skillful newcomers. Compared to earlier models, engineering improvements have reconfigured and redesigned racquets that greatly enhance anybody's game. Now made of high-tech lighter, stronger, springier materials, they make smashes more likely. Since bigger racquets make it easier to make contact with the ball, racquet sizes have increased to enormous dimensions—72 square inches in 1976; 110 in 1997; and 137 in 1996.[5] Size and expanded sweet spots are changing the speed of tennis. All told, hi-tech improvements will increasingly make (almost) everybody resemble an "ace."

Less acceptable, but still likely, are new ways for amateurs to use equipment to get an edge, e.g., Big Bertha golf clubs, now confined to amateur use, have a special trampoline-like recoil advantage and extra-larger and extra-heavier heads. They make it easier to hit golf balls and drive them much farther. In addition, the doping of the face of golf clubs with goop to minimize slicing will probably persist. Aerodynamic dimpling of golf balls to lengthen drives will probably also continue.[6]

Organized sports, by the way, can be expected to constantly adjust to minimize technological advances that provide an edge to amateurs. Basketball has become dominated by literal giants. Slam dunks and "stuffs" are becoming so common, and long-distance shots so unerringly accurate, that dimensions of the target are on the verge of change. Current basket height of 120 inches soon may be replaced by hoops placed 130–140 inches above the playing floor. Backboard width, currently fifty-four inches, may be replaced by ones measuring forty-five-to-fifty-inches wide. Goal posts—for hockey and soccer—may get smaller. Football ground games may give away to talented toes that can boot field goals from fifty to sixty yards out. To add more excitement to the competitive battle on the field, the current goalpost width of twenty-four feet might be reduced to somewhere between twelve to twenty feet. Soccer fields—which once were four miles long (since 1850, a standard 1,000 yards)—may, once again, lengthen.[7]

ADVENTURE SEEKING

Thrill-seeking activities have been and always will be an important aspect of a younger person's lifetime experiences. An exciting and ever-increasing array of activities enable participants to improve skills, gain a sense of mastery, enjoy the sheer exhilaration of peak experiences, emulate the accomplishments of personal heroes, and break up humdrum everyday routines. Adventuresome thrill-seeking pursuits will undoubtedly grow in popularity, as individuals continue to seek tests of their capabilities, courage, and go-for-broke "craziness."

Participants and on-lookers at X Games (hot dog skiing, skateboard tricksters, snowboarders) and contests of all kinds (motorcross hill climbing or ice climbing a frozen waterfall, for example) will enjoy cutting-edge performances. (Many feats appear unimaginably difficult. They *are*. Participants push the envelope in taking skills to the outmost limits).

Affluent individuals will have more and more opportunities to experience high-risk recreation options with a pro alongside; e.g., former President George H.W. Bush's skydiving; Dennis Tito's astronaut adventure; "Top Gun" aerial dogfights; white-water rafting; free ballooning; soaring; swimming with dolphins; extreme driving in souped-up stockcars; evasive and stunt driving; etc. Or, they will be able to enjoy a "play alongside" with pros, including golfers, tennis champs, baseball stars, football heroes, and kung fu experts.

Some of us will chose to enjoy the "feeling" and exhilaration of such undertakings in *controlled* environments (or virtual reality set-ups) that minimize risks, e.g., free falling in wind tunnels; surfing man-made indoor waves (adjusted for speed, height, and distance); swimming against the current in at-home small-sized pools; trying ski and surfboard treadmills/simulators; batting robot-pitched baseballs or tennis balls; copiloting a vintage aircraft; and so on. Another variation here involves fantasy camps of all kinds, including pretending to be a spy (vicariously living the life of James Bond); pretending to be a detective (mystery train rides that allows one to come up with a "solution" of a "murder"); pretending to be an extreme outdoorsman (as in a survival training camp); and so on.[8]

ADVENTURE TRAVEL

Another likely big gainer, adventure travel will allow you to experience dog sledding across endless snowfields, scale the tallest peaks, catch the biggest waves, trek the Poles, or even journey into outer space. Not surprisingly, these experiences serve vacationers for whom money is no object. Some exotic options are already available: Entrepreneurs, for example, are taking cash deposits in advance of being able to deliver suborbital weightless flights, and later, longer-term fly-bys and stays in outer space. These bold ventures are likely to attract thrill seekers and those who want to be first to do anything—and everything.

Russia's Space Agency has already sold some eight thousand crew hours aboard its somewhat decrepit International Space Station for $60 million.[9] Zegrahm Expeditions (Seattle, Washington) offers cosmonaut-astronaut training in Star City, Russia, including a parabolic flight allowing participants momentarily to experience zero-gravity weightlessness ($14,900). NASA uses a converted KC135 (military version of a Boeing 707—dubbed the "vomit comet") to arch occupants in parabolic flight providing weightlessness for about twenty-five seconds. Near-space hotels already planned include orbital stays; geostationary positioning; lunar fly-by; moon and planet landing/exploration; and, ultimately, space settlement and colonization opportunities.[10]

Adventure travel company surveys estimate that ten thousand persons would pay $1 million or more for an adventure in space. Advocates suggest two-week vacations could be priced in the $500,000 range per person. Serious promoters, however, concede that orbital payload will have to drop to $100 per pound before significant space tourism takes off—a level they think soon achievable.[11]

ShareSpace, formed in 1998 by former astronaut Buzz Aldrin, estimates at least forty thousand individuals would spend $1 million each to visit space. Plans call for hosting one hundred passengers in clusters of orbiting hotels at a cost of $100,000 per guest. Aldrin calculates launch costs of $2,000 to $10,000 per pound inserted into orbit, the difference depending upon whether U.S. or Chinese rockets are used. Zegrahm Expeditions began taking reservations for a seven-day space flight in October 1997. By early 2000, 250 reservations for the $98,000 flight had been taken.[12]

TOURISM

The world's *largest* employer, tourism is expected to enjoy still greater gain in its numbers as the world's rapidly growing middle-class affluence provides the means for pursuing a huge range of pastime activities, including travel to places people used to only dream about.[13] They may be ushered there on palatial cruise ships, supersonic jets, or rustic windjammers. Facilities that range from dude ranches and bed-and-breakfasts to elaborate luxury hotels in exotic travel destinations will increasingly accommodate virtually any interest or pocketbook. Major gains are also expected in less high-scale but no less rewarding varieties of trips, e.g., eco-tourism, visiting sites made famous by film or history, trekking the holy lands, explicitly seeking romance (on singles-only trips)—the tourism option list is endless and promises only to get evermore colorful.

CRUISE SHIP GETAWAYS

Large and sumptuous as the largest cruise ships may be, they pale in comparison to what is on the drawing boards. The *Freedom Ship*, planned by Norman Nixon's Florida-based Engineering Systems, measures one kilometer in length and rises a majestic twenty-five stories high! So huge is this ship that ninety-eight-foot ocean swells will not affect it. Penthouses aboard go for about $8 million. This floating city will accommodate nearly sixty-five thousand residents and a crew of about fifteen thousand. Plans call for a two-thousand-member security force, a six-hundred-bed hospital, schools (ranging from kindergarten to college), a flight deck atop, and a marina off the stern.[14]

SPAS AND RELAXATION THERAPY

Spas and relaxation therapy resorts are multiplying and expanding their range of appeal. Day spas increased from thirty to sixteen hundred between 1989 and the late-1990s.[15] They offer mud packs, seaweed and herbal wraps, loofah scrubs, massages, pedicures, ultra-violet lamp treatments, aromatherapy, aerobics, and every exercise imaginable. Natural mineral streams, and a full range of beauty, relaxation, massage, and exercise regimes are provided. Liquor, tobacco, saturated fats, salt, and cologne are forbidden, and cell phones and computer use are strictly limited. While most clients are women, this is likely to change as fast-track men increasingly seek refuge and relaxation.

Recognizing the important role recreation and exercise play in well-being and morale, employers are likely to open more and more on-site gyms, wellness centers, and even indoor pools. Catering to such interests, special time-off arrangements may develop before 2010 that allow employees who are serious surfers to leave work when wind conditions are favorable to windsurfing, and the same for skiers, and so on. Look for more and more of these progressive sorts of overdue accommodations in the years ahead—and consider helping to promote them![16]

SUMMARY

Our current fixation on getting things, on shopping without pause, *may* finally be waning. Over the next ten years, we may conclude there MUST be more to life than 24/7 work and Mall mania. A new focus in life may slowly and steadily gain adherents; an antimaterialist focus that includes cultivating enriching recreational experiences. Instead of crassly piling up junk in the basement, closet, and garage, we may find greater-than-ever satisfaction in recreational options.

Only this much seems clear: Recreation choices will play a larger—and more varied—role in our everyday lives out to 2010 and beyond. Keep an open mind and experiment boldly—and yet also cautiously—with some of the exhilarating and empowering opportunities that come your way; many of which we cannot even imagine today.

NOTES

1 John P. Robinson and Geoffrey Godbey, *Time for Life: The Surprising Ways Americans Use Their Time* (University Park, Pa.: Penn State University Press, 1997), 184.

2 Kevin Hand, "*Off the Deep End*," *Newsweek*, 2 June 2003, 86. Not only do participants get an adrenaline rush from challenges, adulation from peers and on-lookers (plus a sense of self-esteem), but also test their innermost mental as well as physical

capabilities. Careening down a raceway just inches off the roadway on a luge at 72 mph provides a mind-blowing thrill and spectacle. Some others take the easy way. Rather than hike up a mountain for half a day, helicopters ferry their skis, bikes, or hiking gear to scenic pinnacles where full-course elaborate fixings for picnics are enjoyed before the relatively effortless run downhill.

3 Scott Boeck and Web Bryant, "Batter Up!" *Fast Company*, April 2000, 382.

4 Steve Kotler, "Adidas ReBoots," *Wired*, June 2002, 38.

5 Joanna Marchant, "Bigger Balls, Please," *New Scientist*, 2 September 2000, 10.

6 Mark Hyman, "Psst! Wanna Buy an Illegal Golf Club?" *Business Week*, 31 July 2000, 38; "What the Heck Is That Stuff? Baby Oil? Ski Wax?" *Popular Science*, June 2002, 47–51.

7 Graham T.T. Molitor, "Here Comes 2015: The Onset of the Leisure Era," *Vital Speeches of the Day*, vol. LXVI, no. 20, 1 August 2000, 625.

8 Graham T.T. Molitor, Unpublished manuscript, March 2004.

9 Peter Spotts, "Not 'Gonna Be a Long, Long Time 'Til You Can Be a Rocket Man," *Christian Science Monitor*, 26 April 2002, 3.

10 Diane Lanigan, "Space: Launch of a Tourist Trade," *Popular Science*, January 2002, 20; Tara Weingarten, "Life in the Fastest Lane: Risky Business," *Newsweek*, 21 February 2000, 60; Alan Farnham, "Personal Space," *Forbes*, 30 November 1998, 404–405; "Full Throttle Thrills," *Popular Science*, May 2003, 97.

11 Greg Klerkx, "The High Life," *New Scientist*, 30 August 2003, 25–28; Kristina Stefanova, "Perfecting Tech-Net," *Insight*, 7 February 2000, 32–33.

12 Dan Cray, "Will We Take Vacations in Space?" *Time*, 10 April 2000, 3.

13 Geoffrey Godbey, *Leisure in Your Life: An Exploration*, 5th ed. (State College, Pa.: Venture Publishing, Inc., 1999), 216. Employment opportunities, by the way, abound. You will be able to work at developing specialized equipment, including daredevil snowboards, free-fall parachutes, "hotdog" skis, killer skateboards, water jet scooters, snowmobiles, and extreme sporting gear of all kinds. In addition, many will be needed to outfit users,

and provide instruction and support. Equipping yourself to try out some new activity can be expensive. Experimenters, however, will not have to invest large sums in special clothing and gear. They will increasingly be able to rent it. Dabblers will not have to wind up with closets full of expensive stuff they no longer have any interest in keeping.

14 "A Life Less Ordinary," *New Scientist*, 25 March 2000, 30–34.

15 Beth Brophy, "A Vacation of One's Own," *Business Week*, 27 March 2000, 242E12–242E14.

16 Graham T.T. Molitor, "The Next 1,000 Years: The 'Big Five' Engines of Economic Growth," *The Futurist*, December 1999, 15. Recognizing that vacation pursuits grow more expensive, some countries and a few U.S. employers may soon dole out bonus pay while on vacation.

REFERENCES

Baker, Peter. "U.S., Russians Agree to Allow 'Space Tourists.'" *Washington Post*, 10 August 2001, A20.

"Bigger Tennis Balls—Speed Limits." *Newsweek*, 1 January 2000, 80.

Boeck, Scott, and Web Bryant. "Batter Up!" *Fast Company*, April 2000, 382.

Brophy, Beth. "A Vacation of One's Own." *Business Week*, 27 March 2000, 242E12–242E14.

Cray, Dan. "Will We Take Vacations in Space?" *Time*, 10 April 2000, 3.

Farnham, Alan. "Personal Space." *Forbes*, 30 November 1998, 404–405.

"Full Throttle Thrills." *Popular Science*, May 2003, 97.

Godbey, Geoffrey. *Leisure in Your Life: An Exploration*. 5th ed. State College, Pa.: Venture Publishing, 1999, 216–218.

Hand, Kevin. "Off the Deep End." *Newsweek*, 2 June 2003, 70.

Hyman, Mark. "Psst! Wanna Buy an Illegal Golf Club?" *Business Week*, 31 July 2000, 125.

Klerkx, Greg. "The High Life." *New Scientist*, 30 August 2003, 25–28.

Kotler, Steve. "Adidas ReBoots." *Wired*, June 2002, 38.

Lanigan, Diane. "Space: Launch of a Tourist Trade." *Popular Science*, January 2002, 53.

"A Life Less Ordinary." *New Scientist*, 25 March 2000, 30–34.

Marchant, Joanna. "Bigger Balls, Please." *New Scientist*, 2 September 2000, 10.

Molitor, Graham T.T. "Here Comes 2015: The Onset of the Leisure Era." *Vital Speeches of the Day*. Vol. LXVI, No. 20, 1 August 2000, 620–631.

_____. "The Next 1,000 Years: The 'Big Five' Engines of Economic Growth." *The Futurist*, December 1999, 13–18.

_____. Unpublished manuscript, March 2004.

Robinson, John P., and Geoffrey Godbey. *Time for Life: The Surprising Ways Americans Use Their Time*. University Park, Pa.: Penn State University Press, 1997.

Spotts, Peter. "Not 'Gonna Be a Long, Long Time 'Til You Can Be a Rocket Man." *Christian Science Monitor*, 26 April 2002, 3.

Stefanova, Kristina. "Perfecting Tech-Net." *Insight*, 7 February 2000, 32–33.

"Surfing Indoors." *The Economist*, 18 December 1993, 86.

Weingarten, Tara. "Life in the Fastest Lane: Risky Business." *Newsweek*, 21 February 2000, 60.

"What the Heck Is That Stuff? Baby Oil? Ski Wax?" *Popular Science*, June 2002, 47–51.

Zorpette, Glenn. "A Taste of Weightlessness." *Scientific American*, November 1999, 26.

Courtesy of The Venus Project
Designed by Jacque Fresco and Roxanne Meadows

■ Essay Ten ■

SPORTS 2010: THE FUTURE IS NOW

Richard F. Gerson, Ph.D.
President, The HEADcoaching Institute/Sports and Performance Psychologist

As sports are representative of a society's strengths *and* weaknesses, what might be the future of sports in the United States? How much appears so bright you'll need shades, and how much may prove only the same old same old?

By 2010, for example, which sports are likely to evolve into something different? Will women have equal performance abilities? (Some women college basketball teams can already outscore some male teams.) Will teenage athletes be paid five- or six-figure salaries to play high school sports? Will pro salaries keep escalating? Questions, in short, abound, and merit consideration, if also provocative forecasts.

THE DEATH OF AMATEUR SPORTS

Look back with me to 2004. LeBron James, a phenomenal nineteen-year-old basketball player, made the leap from high school to the NBA. He was touted as the next Michael Jordan; the future of the league. Yet, before he was all-world in the NBA, he was nominated as the Amateur Athlete of the Year.

Sportscasters made a joke about the nomination saying he was never really an amateur athlete. He always had some sort of sponsorship while he played ball. For example, reports on ESPN in 2003 stated that his high school team was charging other teams $15,000 a game in appearance fees.

Does that sound like amateur sports to you? Other high school teams were on national television. Amateurs? It is scary to think that high school athletes may become paid players;

minor leaguers, if you will, for the pros. That is what happened to college athletes, and now it is trickling down to high school. The fun is gone, unless we put the brakes on.

One scenario might look like this: High school players are passing up their junior and senior years to make themselves eligible for the professional sports drafts. In previous years, you had to graduate high school to be eligible for draft status by the pros. Now, through a series of ongoing court cases, younger and younger players who think they are capable of playing with the pros have challenged the system and won.

At least, they won the legal case. So now they get drafted and use their first four or five years in the pros to really learn the game. That is, of course, if they don't get injured or cut from the team.

What about the high school teams? You can already get college games on pay-per-view or seasonal subscription packages for TV. Now, you are likely to get high school games also. Where will it end? Will the long arm (or is it the lure arm?) of financial payoffs from sports reach down into middle school, little league, and youth football? (As for me, I say—STOP THE MADNESS and let kids be kids).

One organization that is letting kids still be kids is the AAU—Amateur Athletic Union. This wonderful organization is likely to continue hosting championships for true amateurs in a variety of sports. Plus, many professional and Olympic athletes will probably continue to get their start in the AAU. No matter what happens with the "professionalization" of amateur sports, I personally do not see the AAU "going away." It is definitely a much-needed organization for all athletic participants.

DON'T FORGET THE GIRLS (WOMEN)

You may be thinking, while reading this, that the pay-for-play only applies to boys and men and their sports. That is incorrect. The same holds true for women. They will have the same financial advantages as men, although those advantages will still not be as lucrative. Women will have their pro sports, yet they probably still won't get paid as much as men.

The situation in 2010 for women will probably be the same as it was in 2004. Same sports, different payouts. It doesn't matter if it is golf, tennis, basketball, soccer, boxing, pro wrestling, or water polo. The men are still likely to be paid more than the women to play.

Now, I'm not saying that it was right in 2004 or it will be right in 2010. The payouts are still based on the draws, meaning how many people are attending or willing to pay to see an event. The truism is that more people pay more to see the men/boys play than they do the women/girls.

WILL INFORMATION TECHNOLOGY MAKE SPORTS BETTER?

I have to reminisce a little bit. Back in the prehistoric days (1970–1980 or so), quarterbacks called their plays at the line of scrimmage. Coaches sometimes shuttled in other players with the play that should be run.

There was not a wireless headset built into the helmet of the quarterback. The quarterback had to think, plan, and make decisions on his own in a very short period of time. Now, all the quarterback has to do is listen to the coach call the play through a headset and make whatever changes he sees fit at the line. Has technology made football a better game?

In 2004, they still used instant photos of the field taken from the press box to show player alignments. In 2010, they will undoubtedly use digital photography sent wirelessly to a laptop or tablet computer on the field for viewing by the players. There may also be enough "performance algorithms" built into the computers to analyze the actual and projected movements of the players based on future situations. This will enable coaches to better predict the impact of their next set of play calls.

This technology will probably filter down to all levels of sports. It may even be possible to use wireless handhelds or even eye computers to give players instructions, a view of the game from various angles, or to show them how the play should be run before they actually do it. Technology, in short, can be a great addition to sports, but the question fans will ask is, has it taken away from the *true* performance? Or has it made the performance better? Thanks to the new training methods and new gee-whiz technologies, athletes will run faster, jump higher, and play longer. World records will fall on a regular basis.

People will get really pumped up to break the next world record. No matter what the maximum becomes, however, there will always be something beyond it that will push athletes to go beyond their capabilities.

WILL BIOCHIPS MAKE SPORTS BETTER?

While athletes used ergogenic aids, steroids, and performance-enhancing drugs prior to 2004 to give them an edge, in 2010 some are likely to turn to biotechnology. Someone is likely to soon invent, and a league might even then approve, a biochip that can be implanted to help the athlete perform better.

It would be similar to GPS chips in the early twenty-first century that were implanted in pets so their owners could track them, and that some parents have placed in the arms of their children for safety purposes. [See John Cashman's related essay in volume four.]

These sports biochips will be designed to increase muscle strength and endurance, increase speed, and reduce fatigue, to name just a few benefits. Electrical stimulation of muscle groups specific to a sport will create the proper neuromuscular pathways in the nervous system and the brain. This will speed up signal time and make athletes faster and stronger.

SPORTS FOR THE PHYSICALLY CHALLENGED WILL GROW

We are probably going to see an increase in the number of players participating in sporting events for the physically challenged. Track and field, for example, is likely to see world records for the physically challenged breaking the records of "normal" performers from fifty years ago. This will definitely be due to the technology, training, and inclusion of more people in the sport.

Wheelchair basketball is another example of a sport that is

likely to grow, even to having a world championship tournament. While the sport was originally designed for wheelchair athletes, so-called normal athletes have begun playing the sport for the workout, the competition, and the camaraderie. As more and more able-bodied and physically challenged athletes get involved, this sport is likely to see the same level of international explosiveness and exposure as traditional basketball.

ADDITIONAL NEW WAYS TO IMPROVE SPORTS PERFORMANCE

Nutrition, for example, will probably improve, as will training methods. More and more athletes will have personal sport psychologists or mental coaches, much as they have strength coaches or position coaches on a team. The mental aspect will become even more important because the training methods and technologies will have evened out the playing fields.

EXPANSION IN WORLD CHAMPIONSHIPS

The growth of sports in 2010 will be due to an increase in the desire of people to stay physically active without necessarily having to participate in an exercise regimen. As more and more people play the games, we must accommodate them by developing more leagues, more outlets, and more "championships."

In the United States, we are all familiar with the Super Bowl (football), the World Series (baseball), March Madness (NCAA basketball), the NBA Championships (professional basketball), the Majors in golf, the Grand Slam in tennis, and the World Cup (soccer). We are likely to see an expansion of all these sports to include more international competition (a true World Series, for example).

We may also witness world championships in minor sports, such as lacrosse, table tennis, water polo, etc. And we will always have the Olympics. So, at the highest level of sports competitions, we probably will have more championship games and tournaments.

And, at the amateur level, we will definitely see a similar effect. In the United States, the AAU will probably continue to grow its participation base, as will the amateur sports organizations throughout the world. Eventually, amateur athletes should be able to compete against each other in all the major and most "minor" sports, even more than they do now.

COMPETITION PERSISTS

You may practice a sport without keeping score, but you'll never play against a competitor without trying to win. Ask any athlete what their goal is and they will tell you they want to be the best. The thought that it is just a game is nice, but that is all it is anymore—a *thought*.

And that may not be so bad because most aspects of life require competition. Even though there are more losers than winners in sports, business, and life, that does not mean people will give up keeping score and working harder to get better. Done properly, scoring can be a great motivator.

The thought of sports becoming more cooperative is still a nice concept, but it is unrealistic by 2010. Competitive pressures to achieve and succeed continue to grow. Most people would not participate unless they could keep score against an opponent or themselves. While it is nice to say that sports should be cooperative, it is more appropriate to say that sports help everyone strive for and eventually feel like a winner.

Athletes are the epitome of people who want to win. The push in the late twentieth and early twenty-first centuries for more cooperative sports and games has died down. There are still games of these types, but they are used more to build internal team spirit or group morale in corporations. The ultimate, nonscoring game was T-ball in little league. Yet many parents kept score even if their kids didn't and the rules said you shouldn't. These folks probably passed the scoring bug on to many of their children.

SPORTS HYBRIDS SUCCEED

People have tried to combine sports for years. Movies like *Rollerball* (1975, 2002) combined roller derby, hockey, and motorcycle racing into a sport that really got the viewers involved. Water basketball, a recreation game that involved swimming, water polo, and basketball, was played in country clubs and swim clubs throughout the United States. Slam Ball, which was glorified and promoted on TV in 2004, was a combination of basketball, trampolining, and the body-checking from hockey.

Other hybrid sports that had a small following in 2004 will maintain that small following in 2010. The Ultimate Fighting Championships, in its various forms, along with the independent professional wrestling associations that combine wrestling with who-knows-what, should continue to be part of the sporting landscape. However, they will never dominate the airwaves like professional boxing or professional wrestling (in prime time).

How well did these hybrid sports succeed and how well can they succeed? They will always have a place in the hearts and minds of a small number of followers and players. They are a nice diversion from traditional sports, but they will never take the place of the more traditional sports.

The legacy of traditional sports makes it very difficult for hybrid sports to establish a strong foothold and following. That does not mean we shouldn't try to play or invent new games. They can be a lot of fun!

A NEW PHILOSOPHY OF SPORTS

What we should be doing in 2010 is the same thing that should have been done all along. We should be encouraging people of all ages, especially middle and high school students, to participate in as many sports as possible.

Participation will help in numerous ways: Players will learn athletic and movement skills; communication and collaboration skills; and social and emotional skills. They will also learn that they can win by helping others win and they will find out that they are either a "great" athlete or that they were cut out for coaching/teaching/helping others learn.

A NEW PSYCHOLOGY OF SPORTS

By 2010, sport psychology will have fully emerged from academia. We saw the beginnings of this in 2004 as many sport psychologists began working with nonathletes, corporations, and organizations to help them improve their performances. In fact, the sport psychologist has evolved into the sports *and* performance psychologist.

The new psychology of sports, or maybe FOR sports, will be a focus on the mental, emotional, and psychological gains and strains of sports participation. Coaches will be trained in the basic techniques of sport psychology. As athletes move from middle school to high school, college, and beyond, they will meet regularly with sport psychologists for both discussion/coaching sessions and mental training sessions.

College and pro teams will all have a sport psychologist on staff or affiliated with them to help mentally train and condition the athletes' minds. The stigma of working with a sport psychologist will be removed as players and coaches realize that it is not just about therapy but more about performance enhancement.

Here is where everyone can receive a tremendous benefit. Throughout the years, sports coaches have motivated their players. Players have motivated themselves. Yet who takes care of the emotional burdens brought on by sports? Who helps athletes, regardless of their age or skill level, when they must transition from sports to the "real world?" How and when is the athlete taught to control his or her emotions in a social setting when they were always allowed to express themselves during their games?

SUMMARY

From now through 2010, our goal will undoubtedly remain as it is at present—to maximize the benefits every one of us, duffer or pro, can receive from sports, whether as onlooker or player.

While details tomorrow will undoubtedly differ, sports will always be with us in some (probably increasingly exotic) form. For as far as we can forecast, the physical, psychological, emotional, social, and achievement aspects of sports will uniquely meet a deep-set need as can nothing else.

FURTHER READING

Barkley, Charles. *I May Be Wrong But I Doubt It*. New York: Random House, 2002. This book was written by one of the top 50 NBA players of all time. It is an open and honest account of Barkley's thoughts on a variety of topics, much the same as his commentary as a television announcer for basketball.

Brown, Jim. *Sports Talent*. Champaign, Ill.: Human Kinetics, 2001. This book describes how to select and develop talent in athletes. It is useful for both coaches and administrators, as well as young athletes who are trying to improve their skills.

Dorfman, H.A. *Coaching the Mental Game*. Lanham, Md.: Taylor Trade Publishing, 2003. The author is a recognized authority in sport psychology for baseball players. This book teaches a variety of sport psychology techniques that are applicable to athletes of all ages and is presented in an alphabetical order format making it an easy-to-use reference.

Gerson, Richard. *HEADcoaching: Mental Training for Peak Performance*. Indianapolis, Ind.: 1st Books Library, 2004. This book teaches multiple mental training techniques for athletes and students to use to achieve peak performance, perform well under pressure, and to train their brains to work more effectively. It is based on more than twenty years of research and experience in the fields of sports and performance psychology.

Thompson, Jim. *The Double-Goal Coach*. New York: Quill, 2003. The author clearly explains why coaches should always have two goals in mind for youth athletes. One is related to the outcome of the game and the other is related to their own self-development. The author is also the founder of the Positive Coaching Alliance.

Youth Sports Trust. *Young Athletes Handbook*. Champaign, Ill.: Human Kinetics, 2001. This handbook covers many aspects of being a youth athlete, including what it takes to succeed in sports and other areas of life. The book also is a great reference source for parents and youth coaches.

WEB SITES

www.aausports.org
The main site for the premier amateur athlete organization.
www.cbssportsline.com
CBS television's sports-related site, providing scores and stories.
www.cnnsi.com and www.si.com
Sports Illustrated's site—need we say more?
www.espn.com
The home page for TV's number one sports channel.

Professional Sports Leagues

www.fifa.com
International soccer
www.mlb.com
Major league baseball
www.nascar.com
Professional race car driving
www.nba.com
Professional basketball
www.nfl.com
Professional football
www.pga.com & www.lpga.com
Professional golf
www.soccer.com
Soccer resources
www.usta.com
Professional tennis

■ Essay Eleven ■

TEENS AND TRAVEL TOMORROW: THE BEST IS AHEAD

Marvin Cetron, Ph.D.
President, Forecasting International, Ltd.

As you know, teenagers love to travel, and are among the world's most adventurous and eager nomads. All the more reason to imagine together some of the remarkable changes likely in the years immediately ahead, changes rooted in today's realities. A majority deal with new technologies, for an obvious reason: Technology changes a lot faster than do social factors, and it is often the catalyst that eventually forces societal change.

Our dates, of course, are "guesstimates"; they could be a few years off in either direction. However, we at FI will be surprised if many of these forecasts fail to materialize. And when we are wrong, it is likely because even bigger (often impossible to anticipate) changes made them obsolete. When forecasters peer far out—say, beyond the next ten years—our mistakes are usually failures of *imagination*. The far future is always stranger than we commonly anticipate (hence our appreciation for the best of science fiction).

2010

Invisible Idiot. That is how one early language-translation program, converting from English to Russian and back again, interpreted the phrase, "Out of sight, out of mind." The software available on-line today does not do much better, as anyone who has tried to read foreign-language e-mail can attest.

But within a few years, well-equipped tourists are likely to carry pocket computers capable of translating idiomatic speech from any of half a dozen major languages into any of the others, in real time. Just talk into the box and that Parisian waiter will know exactly what you want. Whether he likes your accent well enough to bring it is another matter.

Who Are You? In the age of terrorism, governments want to know for sure. During this period, probably by 2012, passports and visas are likely to be replaced by biometric identity cards that carry records of your fingerprints, retinal blood vessels, and other permanent, unique proof that you are really you.

Watch the Birdie. Or not. If you can see it, the tiny digital camera built into your sunglasses will capture an image for your friends back home. Expect basic VGA resolution by 2008, higher quality snaps a couple of years later.

Condo Cruising. The first cruise ship with apartments owned by the passengers (price tags from about $900,000) is already sailing from one luxury destination to another, having stopped at the 2004 Olympics and with other major tourist events planned for the future. How quickly other ships follow its lead depends on the global economy, but FI expects that half a dozen of these ultimate RVs will take to the water by 2012.

Superclothes. For adventurous vacationers headed to deserts, mountains, and other hostile areas—one of the hottest areas in tourism, sometimes literally—industrial-strength couturiers are developing "active" attire that cools or warms the wearer, as needed, and collects and stores solar energy to keep your GPS going without heavy batteries. Look for them at trendy outfitters around 2010.

Altered Airports. Those self-service ticket kiosks appearing in major airports will proliferate rapidly and spread to train stations, sports arenas, concert venues, and other transportation hubs and destination. As a result, airport personnel rosters may shrink by 20 percent no later than 2010.

At the same time, radio ID tags will guarantee that luggage arrives where its owners do. And to compensate for the loss of ticket agents, who now act as a first screen for possible terrorists, airport security departments are likely to grow by 10 percent.

See the World—from Above. The X-Prize competition is offering $10 million to the first private team that sends three people on a suborbital junket into space and manages to repeat the feat within two weeks. At least three entrants have made atmospheric test flights, and one is almost certain to make near-space flight practical within the next two years.

After that, it will take less than a decade to build the first large-scale space tourism industry. Seats will probably sell for a lot less than the $20 million paid by Dennis Tito and Mark Shuttleworth to visit the International Space Station. The hardest part is likely to be obtaining insurance for the flights.

Generations of Entrepreneurs. Throughout the world, people age forty and under are starting businesses at a record rate. Among the younger Dot-commers, twice as many say they would prefer to own a business rather than be a top executive, and five times more would prefer to own a business rather than hold a key position in politics or government.

Many of the companies they start will be in the hospitality industry. Expect a huge wave of new resorts, restaurants, tour operators, and other travel services in the next two decades.

End of Immigration. Travel, and almost every land will open its arms to you. Threaten to stay, and you'll be as welcome as a two-cent tip. Throughout the industrialized world, native citizens complain that guest workers and other immigrants are taking jobs, soaking up public resources, and refusing to integrate into the local culture. And though many of those jobs are positions that no one else would willingly accept, there is enough truth in the other complaints to make this movement a powerful political force.

Add post-9/11 security concerns, and it is all but inevitable. Under tomorrow's immigration policies, "your tired, your poor, your huddled masses yearning to breathe free" can darned well stay home.

2020

Faster than a Speeding Bullet Train. The world's fastest trains today operate at a paltry 200 mph or so, though magnetic levitation trains running on closed courses have topped 300 mph. By 2020, the first 500-mph maglev trains may finally carry tourists around Japan, from Los Angeles to Las Vegas, and along other flat, high-density routes. The technology should be available by 2010 or so, but economic and political problems will probably stall its use for years.

Build It, and They Will Come. A host of tourist attractions have proved it in recent years. These include the sixty-story Burj Al Arab Hotel in Dubai, United Arab Emirates, with a seafood restaurant submerged in the Red Sea; the fabulous new *Bibliotheque* in Alexandria, Egypt; the ice hotels rebuilt each winter in Greenland and Swedish Lapland; and the spectacularly popular London Eye. By 2020, we expect to see at least sixty new destinations built. Most will have some unique appeal, but six of the world's major cities will erect their own versions of the London Eye.

How Do You Spell Kaopectate *in Russian?* Point your camera at a sign or label, and it will tell you what is written there. IBM already is working on the technology to translate text written in Cyrillic, Arabic, Hebrew, and Chinese characters into Western languages. Suddenly, we won't have to speak the local language to identify the drugstore and find what we need. By 2010, automatic translators for written material, probably built into cameras and other digital hardware, will be standard cargo for well-equipped tourists.

Vacation Offer You Can't Refuse. It is a way of life in Europe. Cities empty in the summer as workers head off for a month of R&R that is guaranteed by law. The day may likely come when U.S. workers also enjoy shorter workweeks and mandatory vacations, despite the objections of politically influential employers.

The reason? Jobs lost from manufacturing in recent years are just the beginning. Automation and global competition will continue to squeeze jobs from the U.S. economy until, in the long run, it becomes impossible to create useful work for all who need it. The only answer will be to cut the workweek, add time off, and open new jobs to fill in for vacationing workers. The result will be a burst of growth in tourism like nothing the industry has ever seen.

(Please note this is our most "iffy" forecast. A combination of dramatically reduced birth rates, limited immigration, much better public education, rapid economic growth, and other changes *might* conceivably forestall a job crisis indefinitely.

Yet what we are experiencing is high immigration, unexpectedly high birth rates, generally ineffectual public education, and the threat that exploding government deficits could weaken our economy for years. And it seems likely that the day will come when few noncreative tasks still require human hands.

At some point in the future, much-expanded leisure seems all but inevitable. The late 2020s are our best guess at when this will occur, but no more than a guess.)

Advertising in 3–D. Today, holographic videos are cutting-edge technology. Twenty years from now, they will be consumer products. For tour operators, hotels, and other segments of the hospitality industry, 3–D videos—delivered either by mail or over the Internet—will be the ultimate advertising medium, showing potential visitors exactly what they will experience at their destinations. Of course, in a few years they will be supplanted by virtual reality systems that recreate the vacation experience even more vividly.

UN Uber Alles. The United States' go-it-alone foreign policy is an aberration that cannot last. A global economy calls for global institutions, and that means power will inevitably flow toward the United Nations, the International Court of Justice, and the few other bodies with worldwide jurisdiction. To provide broader representation in those bodies, India and one of the Scandinavian countries will finally be admitted to the UN Security Council. This will gradually provide much more uniform laws, regulations, and standards for the hospitality industry and other multinational businesses.

It's a Gas, Gas, Gas. Technologically, hydrogen-powered cars are just over the horizon. Economically and societally, they will not make the grade until some environmentally minded government mandates a change to hydrogen power. Once that happens, industry will ramp up hydrogen production and put fueling stations in every community.

Then bigger things will happen. By 2025, ecotourists visiting the Arctic, Antarctic, and other pristine destinations may arrive by hydrogen-powered jets that emit only water as their exhaust and avoid contaminating sensitive environments.

Eco-Backlash. Yet having hundreds of tourists trampling the Arctic tundra and other fragile environments will not sit well with committed environmentalists, even if the vacationers do travel by eco-friendly aircraft. Global protests against this perceived despoiling of our common heritage may quickly give rise to stringent limits on the number of tourists who can visit what little true wilderness remains in the world.

2030

Beanstalk to the Stars. The space elevator was first envisioned some forty years ago. It will climb an enormous cable, like Jack up the beanstalk, to a terminal where passengers and cargo can board spacecraft for the trip farther out.

Until recently, this was just a fantasy; there were no materials strong enough to build the cable. Today, so-called carbon nanotubes up to twenty times stronger than steel are approaching mass production, and engineers say a space elevator could be completed within fifteen years.

Unfortunately, economic and political factors probably will double that lead time. According to current estimates, the first space elevator could be built for about $10 billion. Cost for a trip to space would be $200 per pound or less, compared with $40,000 per pound for the Space Shuttle. At that price, the space elevator will make space tourism routine.

2040

Room Service? Today, there is only one hotel under the sea, the two-bedroom Jules Underwater Resort in Key Largo, Florida, and just two restaurants. Four decades ahead, underwater hotels and restaurants may be almost common. Most are likely to appear in shallow water, where sunlight penetrates to illuminate abundant life; the most spectacular will be located on Australia's Great Barrier Reef.

However, at least one small, spartan, and incredibly expensive hotel may provide accommodations more than five miles down, where guests can see forests of giant tube worms, "volcanoes" of hot, mineral-rich water, and luminescent fish swimming past the tiny, foot-thick portholes.

Think It, and They May Not Come After All. Scientists have dreamed for years of building computers that understand our thoughts and send data directly into our brains. But that means a kind of artificial telepathy. Think into your computer in San Francisco, and someone in Bangalore, India, will "hear" the thought over the Internet.

This offers the ultimate virtual reality. If one person swims in the sea, walks on the moon, or runs a three-minute mile, the rest of us can share the experience from the comfort of our own living rooms. It brings up the obvious question, *why leave home at all?* One answer is snob appeal. Affordable virtual reality will be good enough for many, but the rich will display their wealth by taking the time, and spending the money to go in person—and others will always prefer to feel the breeze on their face.

You Too Can Be the Man in the Moon. Or at least on it. The first permanent moon base is likely to appear in the 2030s. A decade later, it will be capable of accommodating up to 350 people, including 50 tourists. Thanks to a growing array of space elevators circling Earth's equator, a lunar jaunt will even be relatively affordable. As demand grows, a space elevator on the moon could bring the price within reach of solidly middle-class families.

2050

One World After All. Five decades ahead, the dollar and euro will probably be supplanted by a single world monetary unit, ending exchange problems forever. Biometric identity cards will be issued soon after birth, and the data stored in banks accessible by any government. This will make it nearly impossible for terrorists and other criminals to move around undetected, but routine tracking of our daily movements will further erode what little is left of the old-fashioned concept of privacy, once so dear to us.

Universal English. Automatic translators will be useful for only forty years or so. English already is the *de facto* language of business, as French once was the language of diplomacy. By 2050, 90 percent of the people in the world will probably speak English, at least as a second language. In major tourist destinations, the number will be even higher.

2060

Jobs Aplenty. Today, an estimated 14 percent of the world's people work in the hospitality industry. Tomorrow, it may be 25 percent. In part, we will owe this dramatic growth to the explosion of leisure time when shorter workweeks and forced vacations spread from Europe to the rest of the world. But this also is in the nature of the industry. Sixty years from now, personal service could be the only job category that still requires human workers.

Meet and Greet. With a global Internet, lifelike virtual reality, and even computerized telepathy available to all, who needs in-person meetings? Nearly everyone, as it turns out.

Full-contact telecommunications will do for routine conversations, but to meet new business associates, conduct difficult negotiations, or just build relationships over a round of golf, people will need to "press the flesh" for many decades to come. The meetings and expositions segment will continue to struggle with economic, social, and technological issues as far into the future as the eye can see. But there may be more corporate and industry-wide meetings in 2060 than in 2006.

2075

Water, Water Everywhere. At least in the low places of the world. At the rate things are going, global warming will raise the seas by two to three feet in the next seventy-five years. That will mean hard times for lands like Bangladesh and the Louisiana lowlands, which are barely above sea level even now and are sinking even as the water rises.

It will also modify our travel habits, as temperatures and rainfall patterns change. Expect what is left of Florida to turn into baking jungle, while crops bloom in parts of Canada and Siberia that today hold little more than ice.

Atoms in Space. Nuclear power is banned from space by international treaty. Nonetheless, by 2075 long-range space tugs powered by nuclear reactors may be ferrying cargo and very patient tourists out to Mars, the asteroid belt, and even beyond. Travel time: about three months each way for Mars; more than a year for Jupiter.

Oldies but Goodies. Some of the hottest destinations tomorrow would probably be familiar to today's travel agents. As we have seen, fast, overwhelming technological change will undoubtedly bring a host of new options for tourism. However, it also may strengthen our taste for old, familiar things and our need to reconnect with the past. Crowds will still surround the Taj Mahal, the Great Wall of China, and the pyramids; visit the Grand Canyon and Old Faithful; tour the fjords, the Yangtze River dam, and the Amazon; and throng the halls of the Hermitage and the Acropolis.

SUMMARY

Forget for a moment that you are now in high school—and close your eyes and imagine your next vacation, not as it might be this coming summer, but as it might be in 2075. I'm not kidding—we CAN imagine that NOW, and actually learn something of value from the exercise. Come along and see.

In early July 2075, you and your family ride the space elevator to a port terminal sixty-two thousand miles over the equator. There you join nearly two thousand other budget-conscious tourists on an interstellar cruiser the size of an ocean liner. After a stately embarkation, you sail out through interplanetary space, past the giant storms of Jupiter, close enough to Saturn to walk on its rings, and on into the inky blackness beyond. Just four luxurious days later, you arrive at Alpha Centauri, the star closest to our own sun, tour a small but spectacular system of planets, and get out at ... Er, well, probably at a theme park in Orlando, where the whole journey took place in a few hours of virtual reality. Alas, the smart money in physics is still betting that Albert Einstein was right about our not going faster than the speed of light. We may never travel to the stars unless and until we learn to hibernate and spend centuries en route. $E=mc^2$: It's not just a good idea, it's the law.

Nonetheless, many other VERY exciting changes are likely to come to the hospitality industry over the next few decades. While not as dramatic as "Buck Rogers" star travel, they are as important NOW to you and me. The possibilities highlighted above give us a better notion of what is possible—and what we might soon promote (or resist) with our "votes" (our spending dollars) in tomorrow's enormous adventure and travel market. Tighten your seat belts—the best is still to come!

Dan Wasserman © 2004
Tribune Media Services. Reprinted with permission.

■ Essay Twelve ■

THE PLACES YOU'LL GO: TRAVEL IN THE TWENTY-FIRST CENTURY

Josh Calder
Futurist, Social Technologies consulting firm

Been there? Done that? If all your travel options seem unexciting, just wait. As a young person at the beginning of the twenty-first century, you'll have some travel opportunities that no one has had before.

Not all these things will happen by these dates. They are possibilities, not certainties. But your travel dollars, and maybe even your entrepreneurial skills, will help determine whether they do.

There will be huge opportunities in creating new experiences for tourists: Travel will be one of the largest industries of the twenty-first century, with hundreds of millions of new tourists from places like China, India, and Russia.

2012: *TITANIC* TOURS

You wanted to be among the first to see the famed Titanic *with your own eyes, and now it was happening. There it was, looming out of the dark like a cliff.*

The approach over the muddy ocean bottom had hinted at what was to come. The seabed was strewn with the debris of the ship's death: coal, elegant china, and more poignant remains, like the head of a porcelain doll.

As the powerful lights of the sub illuminated the upper decks, the tour leader pointed out each feature, interspersed with old photos and computer simulations on the view screen. It brought stunningly home what had happened here exactly a century ago.

Eeriest of all were the portholes with lights. No matter that the company put them there for illustrative purposes—the lights of the Titanic *still seemed to burn.*

Tourist submarines are spreading. The next step could be deep-diving subs, to explore the great wrecks and undersea volcanoes that were impossible to reach until recently.

2015: AIRSHIP SAFARI

The Tanzanian savanna drifts by below. It is the peaceful pace of this cruising, and the silence, that you like most. There is a slight vibration from the engines but no sound. The animals below don't even seem to notice, perhaps mistaking the airship for a cloud.

The zebras and wildebeests stream over the plain like ants. Twice, the herds bolted, holes opening in them like water around a drop of oil, signaling lions on the hunt. When the airship circled low, you saw all too closely that a zebra had not escaped.

Lunch flying over the salt lake was memorable. At the little tables on the main observation deck, you watched the massed flamingos below, like pink petals in a pool, taking flight in elegant swirls.

Returning each night to the lodge for dinner and bed are pleasant enough, but you want to come back next year, when a bigger airship will include sleeping accommodations. Meanwhile, you want to look into dirigible whale watching in Baja California.

New dirigible designs are taking use of advanced lightweight materials and improved motors. Nonflammable, sturdy, and maneuverable, they may overcome the safety concerns that have dogged airships since the *Hindenburg*. Some people are already discussing pleasure cruises.

2016: UNDER THE CARIBBEAN

It might be easier to take the dolphins watching you through the glass wall if they didn't have that smirk. Sometimes they look like they know that you're changing clothes.

Otherwise, the privacy is unbroken—diving areas are all far from the suites. The vista takes in an unending parade of sea life: schools of wheeling fish of every color, sharks, even a majestic manta ray.

Nights brought unexpected beauty. Leaving the shades open, you were treated to the sea in moonlight, the palest, shimmering shafts filtering down through the waves, revealing fish like ghosts. With the sounds of the sea piped in, there seems to be no wall between you and the water as you fall asleep.

It is worth getting up at dawn for breakfast under the glass dome of the main hall. The gray light slowly transforms into brilliant beams refracted onto the plants and fountain. Later, scuba amidst the sharks. For now, sitting in the light is just perfect.

Seventy percent of the world is under the oceans. This vast underwater realm has its own wildlife and scenery, along with countless shipwrecks and even ruins. Diving is a global industry, but it is only the beginning of possibilities. Basic underwater lodgings are being attempted, but no full-scale projects have been built.

2025: ORBITAL HOTEL

You knew you weren't in Kansas anymore when you went to the zero-gravity swimming pool. Not a pool but a round, soft-walled room, it had just enough water to fill the air with globules of all sizes, from blueberries to quivering boulders, through which "swimmers" gleefully drifted, their collisions continually redividing the silvery orbs.

Thank goodness the dining areas are in the rotating section, where gravity seemed normal enough. Zero-G soup would not be pleasant.

What a view. The Earth fills the glass wall of the dining room in an ever-changing mural: the turquoise of the Bahamas, the snaking dragon shapes of snow-covered Himalayas, and the play of lighting in thunderstorms as you pass over the night side.

You soon knew where everyone was from. "Sydney! There's Sydney!" or "Seattle!" they'd shout, as home passed below.

After lunch, a nap. A zero-gravity room might be fun when you come back with your girlfriend, but for now you prefer to sink into your bed, rather than be strapped to it. Afterwards, it's off to the gym, where you'll put on your helmet and fly.

A Japanese company is already contemplating an orbital hotel. The primary obstacle is the lack of a reliable and relatively cheap way of getting things into orbit. Planned space planes may provide this—but don't expect any bargains.

2026: THE BUFFALO ARE BACK

It is hard to believe it is safe. The sea of shaggy heads, short horns, and humped backs stretches away in every direction, and the grounded hover-rover shook with their passing.

But the buffalo—bison, more properly, as the guide reminded you—seem indifferent to the vehicle, ignoring it as they jostle by. It is a scene out of Africa, but this is Nebraska. Joining the other tourists on the roof, you activate your cam-pin. You want to be sure that your friends back home see this, and also hear the sounds of a thunderous mix of hooves, snorting, and lowing.

Tomorrow, you plan to attend a Sioux buffalo hunt. Performed as a cultural rite, the hunt can be observed by outsiders from the park airships. The fee is hefty, but this is too unique to pass up.

Before they were almost wiped out by American settlers, tens of millions of bison roamed from the Appalachians to the Rockies. Herds were so big that it is said they would have been visible from orbit.

People have discussed recreating some of the herds. It may happen by simple economics: Portions of the Great Plains may become more valuable as wildlife tourist attractions than as farms or rangeland.

2031: PLEISTOCENE PARK

"Hey mom, did you have those as pets when you were little?" you ask. She gives you a mock glare, but she knows you're just kidding: The last people to see these animals lived thousands of years ago. Now, you have another chance.

The mastodon family at the edge of the lake is so close you can see the grass they're eating, and the way the baby grabs its mom's long red hair with its trunk, pulling playfully until she swishes him with her tail. They aren't called mammoths for nothing: Shaggy, with massive curving tusks, they seem larger than any elephant you've ever seen.

Still, your favorites are the giant ground sloths, which resemble a sloth crossed with a grizzly, on steroids. They look pretty dangerous, but their huge claws are only for stripping leaves from trees. Rearing up on their hind legs, they bend thirty-foot trees to eat from the top-most branches.

The moas from New Zealand were nice, but the aepyornises put them to shame. From Madagascar, these giant birds stand ten feet tall, and you saw one guarding an egg the size of a football.

The dodo birds don't do much, but seeing them seemed obligatory, and you loved the sheer ugly awkwardness of them.

Coming next year, saber-toothed tigers, and the giant wooly rhinoceros.

The dinosaurs recreated from DNA in *Jurassic Park* were fiction and may remain so. But more recently extinct creatures, from the Pleistocene era to modern times, may be revivable. DNA samples are available from many extinct animals, not least among them mastodons. The chief difficulty may be finding mothers to carry embryos of ancient creatures, but use of related species and artificial wombs and eggs may take care of this.

2045: THE MOON

"How clear the air is here," you find yourself thinking. Then you remember again that there is no air. The mountains on the horizon are perfectly sharp; you can't tell how far away anything is.

That is part of the majesty of the place. It is the ultimate desert, utterly austere. It is not for everyone, but it has the stark beauty of snowscapes and sand dunes, in gray against a black sky.

And the trip from Earth might be worth it for the view of our home planet alone. Hanging blue and white over the lunar mountains, Earth seems too small to contain everything and everyone you had known before this trip.

Here, "leave nothing but footprints" is not a useful slogan: Without wind, rain, or plants, footprints will remain for millions of years. At the Apollo 11 site yesterday, the seventy-six-year-old footprints of those U.S. astronauts looked like they'd been made the day before. Near those historic locations, they are quite stern about keeping you on the walkways. At least you get to leave a permanent sign of your visit on the Footprint Plot; maybe your great grandkids will see them there.

Tomorrow: golf on the driving range by the hotel. In the weak lunar gravity, anybody can hit the ball six hundred yards.

FURTHER READING

Dirigible Tourism

http://www.deltawillis.com/airship.htm

Airships over Africa—A company that seeks to use dirigibles in tourism.

Underwater Tourism

www.jul.com

The Web site of a basic undersea hotel, Jules' Undersea Lodge.

Bison Watching

www.gprc.org/Buffalo_Commons.html

A site advocating the idea of "the buffalo commons," a large area of the American plains returned to a more natural state.

Recreating Extinct Animals

www.suite101.com/article.cfm/4003/69777

"Cloning Extinct Animals"—an essay on cloning mammoths by a paleontologist.

www.salon.com/tech/feature/2002/01/03/mammoth

"A Mammoth Undertaking"—an interview with a proponent of mammoth cloning.

www.wired.com/news/technology/0,1282,52959,00.html

"Will Tasmanian Tiger Clone Work?"—an effort to clone the recently extinct Tasmanian tiger.

Rob DeSalle and David Lindley, *The Science of Jurassic Park and the Lost World* (New York: Basic Books, 1997).

The science behind recreating extinct creatures.

Orbital Hotel

www.spacefuture.com/tourism/tourism.shtml

A site about the prospects for tourism in space.

www.spacefuture.com/habitat/hoteldesign.shtml

A site about orbital hotels.

Visiting the Moon

www.space.com

An authoritative site offering abundant news and background information on space exploration.

http://nssdc.gsfc.nasa.gov/planetary/planets/moonpage.html

NASA's main page for information on the Moon.

Carl Koppeschaar, *Moon Handbook: A 21st Century Travel Guide* (Emeryville, Calif.: Moon Publications, 1995).

A guidebook for future tourists.

"If there are sixteen moons, it must be Jupiter."

Epilogue

ON USING FUTURISTICS

... teenagers are the future of the nation
... a force to be reckoned with
... a segment seeking to influence the world
based on the moral and spiritual foundation
they have already adopted.
—George Barna,
Real Teens

The better to turn leads from the essays to advantage we close this third volume of four with a sound call to action, a bracing reminder that responsibly advanced forecasts are only useful if they inform actual decisions and actions of yours. Information, the following essay counsels, must become "actionable intelligence" if it is to help as it might.—Editor

■ Essay Thirteen ■

GRASPING THE REINS OF CHANGE

Arnold Brown
Chairman, Weiner, Edrich, Brown, Inc.

In Homer's *Iliad*, Cassandra, the beautiful daughter of King Priam of Troy, caught the eye of Apollo, who wooed her by giving her the power to see the future. But when she spurned him, Apollo made sure the other Trojans wouldn't believe Cassandra's prophecies.

As with much of ancient Greek mythology, the Cassandra story is a metaphor. It tells us that having good information about the future—being right—is not enough. It is necessary that you know *how to think about it* in order to know how to respond to it effectively.[1]

New York Times columnist Thomas Friedman, in commenting about the criticism of the U.S. intelligence apparatus after 9/11, wrote that what occurred was a failure not of intelligence but of imagination. We had information; we had good information. We just couldn't see how to think about it in the most effective way. As novelist John le Carré has written, "The greatest spies are worth nothing if their intelligence is not wisely used."

You can make better use of both imagination and intelligence alike by looking at three trends now decisively changing our world. If you really see these changes for what they are, you can plan for their impacts or you can work to alter their impacts.

TREND NO. 1: INTERNET'S IMPACTS

Plainly, the Internet is shaping a novel future. Your car dashboard or cell phone, for example, may soon read your lips and observe and act upon what you say. Already, devices that can monitor and interact with or report on your every move can be installed in your vehicle.

In the world of work, silent office software—by scanning employees' electronic address books, instant messages, buddy lists, electronic calendars, and e-mail correspondence—can now mine your list of acquaintances for new business prospects—with or without your knowledge or permission. Control over relationships slides in this way away from you, and can be stealthily absorbed into the knowledge membrane (the computer infrastructure) of the organization. Snooping devices can cloud individual autonomy—if we all sit by and do nothing.

The Internet, which can take, can also give—as in inconspicuously bestowing power in ways that challenge the established order. In the *new* new world order, small countries like Norway take on unprecedented military roles and power. Size no longer matters as much as does speed and the ability to bring narrowly defined skills, like mine clearing and mountain reconnaissance, to the front lines.

The Czech Republic, for example, now specializes in detection of chemical and biological agents; Latvia, in live ordnance disposal. Norway is now second to the United States in NATO per-capita spending on defense. These small countries are knitting a blanket of military power that will soon dwarf some of the larger historic players. Acting together, they are creating a new layer of military might previously unseen in modern history.[2]

TREND NO. 2: RISE OF SECULAR FUNDAMENTALISM

A hard-nosed, militant nonreligious mindset now matches the church fundamentalism so rampant in certain religions. To some extent, this phenomenon is fueled by disappointment, by a growing gap between what people expect and what they get—*grating expectations*. That gap leads to discontent, unhappiness, and the kind of anger that can result in deadly militancy and blind fundamentalism.

Resentments that might have festered in dark corners can be gathered together through the Internet to create both engagement and clout. *Darknets* are created by activists to bring together people of like mind in order to create "smart mobs" that can both affect policy and reinforce commitment.

Much of this reflects the fears of people who see our world changing rapidly and powerfully, and who see themselves adrift and helpless. As change continues in its headlong pace, and as new technologies drive many institutions to obsolescence, it is likely the attractions of secular fundamentalist certainty will increase—and pose serious hazard to any who seem to benefit from deliberate change (like you and me).

TREND NO. 3: TRANSMUTATION POSSIBILITIES

For centuries, scientists and pseudo-scientists alike dreamed of using alchemy to transform base substances, like lead, into valuable ones, like gold. It never worked. But now science may have the tools to realize a twenty-first-century version of the alchemist's dream.

We may soon be able to accomplish *transmutation*. That is, turn elements and materials into something entirely different. By changing a material's atomic structure, which nanotechnology makes possible, that material can be transformed into something else, with new properties, some of which have never before been seen in nature. Some physicists have even already created a new form of life—globs of gaseous plasma that, like any other life-form can grow, replicate, and communicate.

But it is not only inanimate elements and other substances that can be transformed. Human beings can, too. Many scientists are eagerly exploring how people can be transmutated into some superior form of humanity through the convergence of nano-bio-info-cogno technologies.

The hope is not only to improve humanity but to more firmly control human evolution in order to create bodies and brains more durable, easier to repair, and more resistant to disease, stress, and aging. By merging biology and electronics, bioartificial replacement parts for the lungs, pancreas, kidneys, and limbs can be created. Artificial muscles can be made out of electroactive polymers. Scientists are even working on altering our brain, much for the better (and not soon enough).

The promise of improvement for both materials and humans seems boundless. It ranges from neutralizing radioactive wastes to helping to make everybody a superior athlete (as, perhaps, in "Lake Woebegone," where all the children are above average). It looks increasingly likely that modern alchemy can help us find substitutes for scarce materials, overcome threats to the environment, and make almost everything better, stronger, longer lasting, and possibly even more beautiful.

Is there a down side? Of course. We don't know what all the consequences of such revolutionary developments will be. New nanomaterials may well create new dangers to human health and the environment. And the new people we might create may have attributes not anywhere as good as hoped for.

SUMMARY

Albert Einstein, on being told of the first atomic bomb that was dropped on Hiroshima in 1945, said, "Everything has changed—except the way we think." In a world so greatly changing, ineffective and rigid ways of thinking about things won't suffice. Looking at the three developments above, you may respond in ways that are perhaps unavailable to older people—to your folks, neighbors, teachers, and writers like me. Many of us are burdened with accumulated familiarity with the way the world *used to be.*

You can more easily discard tired prejudices and shopworn assumptions, because they are less ingrained in you. And because of this, you can more easily respond with hope to information—like that above—about change. In place of stubborn resistance or foolish denial, you can respond by asking—*What can I do with this development to benefit myself, my institution, and my community?*

Hope is important. Because change is frightening.[3] And, to quote le Carré again, "frightened people never learn."

Knowing how to use information will help you to rise above fear and see, understand, and respond effectively to all the changes around you.

What to do? For openers, you could help create a discussion group in your high school to study these radical developments. Together, you can bat around what they may, could, should, and should not mean for your future. You can also form a peer group through the Internet to extend the dialogue and act to help bring about changes you want.

What matters is not what you presently know, but what you are resolved to soon learn. And what you intend to put to good use. For as Aristotle wrote: "In practical matters, the end is not mere speculative knowledge of what is to be done, but rather the doing of it."

NOTES

1 An abundance of information today has caused many people to overlook the basics of good thinking. People are led to believe that all you need is more information. But information is like raw iron ore. You first have to turn it into something else before you can use it. Automobile companies cannot make cars out of iron; it has to be transformed into steel first. Similarly, you have to take information and change it to *actionable intelligence*—something you can use to make useful and intelligent decisions about your career and your life.

2 On a parallel plane, U.S. courts are increasingly embroiled in international litigation, even though they are not institutionally equipped to conduct foreign relations. Their decisions are casting a wide net over, and seizing power from, traditional diplomatic channels.

3 Since September 11, 2001, the world is a very different place, not just for Americans but for everyone. All of us are suddenly deluged with information about change. And, in a highly competitive, fast-paced and much more dangerous environment, we all have to respond quickly and effectively. If we don't, we will miss the opportunities, and we will find ourselves left behind—

with a diminishing chance of catching up. All signs point to this becoming even more true two, five, ten years from now. The pace of change is accelerating, and new scientific breakthroughs and technological developments are coming ever more rapidly.

Appendix

STUDENT FEEDBACK

Sixteen high school volunteers read more than sixty candidate essays and offered feedback on many (though not on all) that influenced the final selection. Their (anonymous) views below are listed in the order in which they arrived back to me. They join me in hoping this material helps you take more from the essays.

HEART FUTURES: THE FUTURES OF FRIENDSHIP AND ROMANCE—*MARCUS BUSSEY*

1) This essay was kind of corny. It reminded me of the movie *You've Got Mail*; sort of like a far-fetched love story. I do not think teens will be too interested in this type of essay; maybe it just wasn't for me.

2) I would reject this one entirely. It started out weak and a little "cheesy" with the first line of the introduction. Furthermore, the essay does not seem to have any original ideas. It is a mix of the movie *You've Got Mail*, the present identity problems on AOL instant messenger and e-mail, the song *Sunscreen*, and the issue of heterosexuality vs. homosexuality.

Love is a complicated and tricky business, especially when one attempts to capture its mystery and fantasy in the realm of what always seem to be clumsy sentences and vocabulary. This author has not created anything new or intriguing, and in a way has only degraded the meaning of love for the next generation.

3) This essay is the most peculiar I've read so far. I don't have a set opinion as to whether I'd include it in your book or not. I just can't seem to decide whether teens would be interested. I liked it but would other teens?

4) I wouldn't include this essay. I liked the format and the idea, but I think the topic would dissuade some from reading it. Marriage—at least for most of the teens I know—is not really

the most appealing and interesting topic. Many teens would find it hard to relate to—even though it talks about the future of our relationships.

5) I like this revised essay much better than the first one. It is more realistic and easier to follow. I am just not sure if this should be in the book. I do not think teens are too concerned about marriages yet. On the other hand, another part of me says they will like it because of the Internet and how that is changing the way we view everyone and everything.

THE FUTURE OF FASHION—*LINDA BROWN*

1) I really enjoyed reading this essay. There were a large number of surprising facts and interesting thoughts on the future of clothing. A majority of teens would find this essay worth reading. At first it seems as if it would stereotypically only interest girls, because of the underlying theme: fashion. But as the essay quickly explains, it is about a more technical look into the future of our apparel. Obviously this is something that affects us all, and I think many would find another person's detailed perspective quite interesting.

2) This essay is very interesting. It is so weird to see what clothing will be like in the future but cool at the same time. I cannot believe some of this stuff is actually real and out on the markets. I think teens will like this essay.

ROMANCING THE CLONE: THE FUTURE OF THE FAMILY AND RELATED ISSUES—*MARILYN DUDLEY-ROWLEY*

1) This essay is both interesting and easy to read; it introduces new ideas often enough to stay interesting. My only criticism is to try to make each forecast about equal in length—at most one long paragraph. It's more consistent, and it prevents distraction from the overall theme of the essay (family) by long separate ideas. Also, maybe you can combine a few of these? Put them into sections? So many of them are about marriage/the lack there of/different forms of marriage.

2) This essay is very interesting. I think it is an important issue to address with all the same-sex marriages happening right now and the redefinition of marriage. I think teens may like this. It sounds a little too adult to me, though. While reading it, it didn't seem like an essay for teens; it seemed more formal. I think the author should try and put something about teens in there, like maybe kids having boyfriends/girlfriends at an earlier age; or relationships in school.

3) Use this essay, because our families are an integral part of our life and I like the "politics" section.

4) I don't think the author defends her predictions well—maybe if she just concentrated on a few of them. None of her predictions were topics that I hadn't already heard before. I didn't really like the way it was organized either; like a loosely connected list.

5) I would definitely accept this essay. It is an aspect of the future I never really considered: the family structure. Regardless of what horrors Hollywood may show us of a future without family and love, I believe the family unit is everlasting. Furthermore, I agree with pretty much all this essay has to say, at the very least in theory.

I also feel that an essay of this sort should be accepted considering the recent attitude of the world. I cannot open my local paper in the morning without reading three to five editorial columns or vents belittling homosexuals or any other person that does not fit society's "norm." You would think with how far the United States proclaims it has come, we would be able to overlook individual and private differences such as sexual orientation.

6) I enjoyed reading this essay and it would be a good selection for the book. But I have a comment: When she references other essays, it sounds out of place; it would be better if she could make them smoother and less choppy. I really like how she had footnotes at the bottom and the questions to ponder.

THE DEMISE OF ETHNIC DIFFERENCES: CULTURAL IDENTITY AND THE INTERNET—*CALEB ROSADO*

1) This was a good essay. It carried a good point across about what the Internet means and is doing to its users, but it seemed to drag on and kind of repeat what it was saying.

2) I liked this essay; I think it brings up a new look at the topic of diversity through the Internet and sends out a good message. I would include it.

3) I really loved this essay. The points made about the Internet were interesting and well written. I also liked the "Celebrating Diversity or Living Wholeness" section of the essay. It will be a good time when men are appreciated by their peers for being men and not looked down upon because of skin color. The Internet is helping to make that happen and merging mankind as one. I do, however, believe that holding onto one's culture is important. If that disappears, then everyone will be the same. The essay begins and ends strong, and makes many intriguing points.

4) I'm not quite sure whether or not you should include this essay. It's an interesting topic that I've never encountered before, but it didn't hold my interest very long. It's not a topic that is at the top of my list, but it might be for some others.

5) I am not sure if teens will be too interested in this topic. Kids my age are already tolerant of other cultures, and it is not because of the Internet. Schools nowadays have all races in them with numerous amounts of ethnicities in each classroom. I think school plays a more important role in ethnic equality than the Internet.

THE FUTURE OF RELIGION: THE DIVINE IN TIME—*ANN COOMBS*

1) I personally found this essay somewhat lacking when applied to the prompt at hand. You were looking for essays describing the future, and this one certainly talks about a potential future in very vague terms. There is no point and counterpoint, as there is only one side presented. As a teenager, I would have liked to read about the other side of the issue, as it is far more pressing to my generation. We care less

about praying and more about walking on Mars, less about Mass and more about stem cells. In all truthfulness, an essay on religion would do little to strengthen your volumes, as very few people associate it with the future. It will still be there, but people my age find little interest in pondering what form it will take. Those who want it will have it and those who do not, will not. We all will have science, though; that is the essay I want to read.

2) There is an incorrect idea in this essay. "The Vatican, however, does not want to alienate many millions of core supporters in developing countries to satisfy a relatively small number of liberal reformers in the West. So, securing radical reforms will probably take another fifty years because opposition is so strong in the conservative Third World."

This idea is not correct. The Vatican does not allow women to be ordained, not because they don't want to "offend" anyone, or alienate conservatives, but because they do not have the power to do this. No one on Earth does. Christ did not give his bishops the power to ordain women but to ordain men.

This essay implies that women will someday be priests in the Roman Catholic Church. This cannot happen, because if woman are "ordained" the sacraments they give will be invalid. It is not a matter of "women's rights" or suppression of women, it is a matter of the way in which Christ set up his church. He gave the power to offer the Mass and sacraments to men, not to women, and no amount of "wishing" it to be different will change this. Women cannot offer Mass, because if they try to do that, the Mass will be invalid, no matter how many women believe otherwise.

3) With the release of Mel Gibson's movie and other issues that have religion as a backing argument, this essay would be a good one to be included because religion is such a big issue. However, I have a hard time thinking that a majority of high school students would take the time to read it, so for that reason I have to say do not include it.

4) I like this essay, but I am not too sure if other teens will. I

do not know if teens will be too interested reading about religion. It started off well by making reference to pop culture, like the movie *The Passion* and the television show *Touched by an Angel*, but then it stopped adding these things. You need to put the pop culture stuff in there to gain the attention of teens, especially with this topic.

5) I liked the essay. It held my attention to the end, and I liked the summary. The information was presented in a succinct way, and I enjoyed the way it described the changes in how people view their religions and how the different religions could change in the future.

6) I absolutely love this essay! I definitely think that it should have a place in one of your books. Everybody can relate to it, whether they are religious or not. It is true that religion is continually a topic for intriguing conversation. Just yesterday, my classmates and I got into a wonderful discussion on religion, and it was truly inspirational. I have always wanted to be a teacher, but for a while now I've been thinking about youth ministries, or maybe the peace corps. This essay is fascinating, and it will be a great source for discussion and everyday conversation. I have no doubt that other high school students will love this essay too.

UPDATING THE TEN COMMANDMENTS: A DECALOGUE FOR THE TWENTY-FIRST CENTURY—*JOSEPH F. COATES*

1) Commandments one and six are clearly objectivist views. They also make sense. The rest are about collectivism, subjectivism, and relativity, which are against the ideals of objectivism. If these are to be a new set of ten commandments, one standpoint has to be chosen, not three. The lead into the commandments was good, but again one idea must be presented; otherwise it gets too jumbled and contradictory.

2) I like this essay. It tells its readers that no matter what changes are thrown at us in the future, we can help control them by following these ten steps.

If the writer could shorten it somehow, it might be more effective. Also I would make the beginning part easier to read.

Before I got to the ten "commandments" I felt myself getting lost.

3) This essay is really good. I like the way it is set up and the style of the piece. It was a really good topic, and I especially like the introduction; it really gets the reader in the mood for the essay.

4) I do not like the part of the first commandment that says, "No hypothetical or imagined powers, including devils and gods, can share responsibility with us." That is very offensive to me. I am a Christian and I do not agree with this statement. It is OK to say that man is responsible but not to say that gods do not share responsibility.

ENTERTAINMENT IN THE NEW LEISURE ERA —*GRAHAM T.T. MOLITOR*

1) This essay, like the other one on recreation, appeals to a majority of students, so there will be a large audience. Also, there are some noted improvements to the original essay.

2) I liked this essay. I thought it had a good topic and was interesting, but it was just too long. It was full of great information, though, and it relates to and could be categorized with his other essay on recreation.

3) I like this essay. It was really cool to see some stats on things we use today and how we will use them in the future. The only criticism I have of this essay is that it only presents facts. It was difficult to digest the whole thing. It needs more breaks between the facts and stats.

PURSUING RECREATION IN THE NEW LEISURE ERA —*GRAHAM T.T. MOLITOR*

1) I don't think the language is particularly interesting and it didn't give a whole lot of information. They were just inferences. Nothing was particularly new or exciting to me. A lot of it was about the advanced technology of today.

2) I enjoyed the unrevised version of this essay, and I like the new and improved essay even more. But the real reason why

you should include this essay is that I can see *many* students reading about sports and recreation in the future because that appeals to a large majority.

3) I really liked this essay. It was interesting and really kept my attention. It provided a lot of new information and made me excited for what lies ahead. Many high school kids would enjoy reading this.

SPORTS 2010: THE FUTURE IS NOW—*RICHARD F. GERSON*

1) The sports essay is good, but I would suggest some elaboration about amateur athletics in the section entitled "The Death of Amateur Sports." Also, it might be useful to mention more about amateur athletics in "Expansion in World Championships."

I myself have participated in the Amateur Athletic Union (AAU) National Championships, and while they might not be well known to the rest of the world, they are the breeding ground for the up-and-coming sports prodigies who are about to break into professional sports. For example, Marion Jones, a famous track athlete, participated in the AAU Nationals before she entered into training for the Olympics.

So, mentioning the current amateur athletic organizations might be useful for athletes who are unaware of their existence.

2) I like this essay. It is very interesting to read about the different sports and how they can change or combine in the future. Almost all of the kids I know play a sport in some way and this essay will interest everybody. However, the author should have included something on the over emphasis people put on sports, for example, the fights between parents at Little League games, dominating parents pushing their children, and maybe even the use of steroids and other illegal substances in the sports world today.

One thing I do not like about this essay is the author's feeling toward the subject. I do not understand why the writer is making sports sound like a bad thing. Maybe he doesn't think so, but that was the feeling I got when I read this essay.

TEENS AND TRAVEL TOMORROW: THE BEST IS AHEAD —*MARVIN CETRON*

1) This essay is very interesting and exciting. I like the way this essay describes each scene so well. I was able to picture each scene with a lot of detail. This is something that high schoolers would love to read and dream to experience in the future—who wouldn't? It is also interesting that at the end of each paragraph the author explains how close we are to having these types of vacations so that the readers understand that it is not as far-fetched as it first may seem.

THE PLACES YOU'LL GO: TRAVEL IN THE TWENTY-FIRST CENTURY —*JOSH CALDER*

1) This essay about future vacations is cool and exciting. It makes you really wonder if the future will bring magnificent destinations like this. I think it is a good essay, but I am not sure if teens will like it. There is a lot of imagery and it may bore some kids.

GRASPING THE REINS OF CHANGE—*ARNOLD BROWN*

1) The essay provides an important message, and I like the way it uses the story of Cassandra in the beginning to draw the readers in and make a metaphor.

2) The trends sound like they want to go toward the ideas of the futuristic movie *Gattica*. The essay was very interesting but those who have seen the movie know what horrors are in store if such things actually happen. The essay is important to include; it proves that not all change is good. In the movie, discrimination was based on your genetic makeup, not your skin tone or nationality—such things that will lead to the destruction of the world. It is important to discuss "the radical developments" in a logical way in order to define which type of change is for the best and which should be avoided. It is important for high school students to learn about this at their present age, because they will be the people who decide the future.

3) I like it. It addresses possible extremes of the future (and

partial present), as well as shows what we can do to fix them. I also like the use of quotes and allusions to the past, showing that this kind of thing has happened before. There is a good blend of future technology possibilities and real-life thinking.

4) This essay was kind of "blah." It did not really do anything for me. It had some good points and good information, but it was kind of boring.

5) I really liked the topic. It is intelligent and imaginative. I thought it was extremely well organized and well presented. The ending really summed up the essay well.

6) I found this essay quite interesting. The fact that it started with a mythology reference did not hurt, as I am a huge fan of all forms of mythology. There was a good mix of history and present-day topics, all in relation to the future. It was a neatly structured transition and did well to use timesets in the form of valid and logical examples.

ABSTRACTS FROM *FUTURE SURVEY*

Michael Marien, editor

The following abstracts were prepared by Michael Marien, founder and editor of Future Survey, *a nonpartisan monthly newsletter published by the World Future Society since 1979.* FS *provides fifty abstracts every month of recent books, reports, and important articles on both global and domestic issues.* FS *carries items on trends, forecasts, and policy proposals on topics such as world futures, the global economy, the Middle East and other regions and nations, security, energy, the environment, governance, education, health, crime, communications, new technologies and their impacts, and methods for shaping a better future.*

WORK/FAMILIES (October 2003)

Families That Work: Policies for Reconciling Parenthood and Employment. Janet C. Gornick (Associate Professor of Political Science, CUNY Graduate Center) and Marcia K. Meyers (Associate Professor of Social Work and Public Affairs, University of Washington). New York: Russell Sage Foundation, August 2003, 392 p.

Seeks to move current policy debates in the United States beyond fragmented conversations that treat child well-being, the conflict between work and family life, and gender equality as separate issues, and beyond the belief that care giving is a wholly private concern. Cross-national comparisons can help move the debate by suggesting new conceptual and practical models for policy design. "*We need to focus our attention on an end vision of what an earning, caring, egalitarian society that promotes the well-being of children might look like the dual-earner/dual-career society.*"

Imagine: 1) a world in which mothers could take a few months away from their jobs following birth or adoption of a child, without sacrificing job security or their paychecks; 2) a world in which both mothers and fathers could spend substantial time at home during their child's first year, while receiving nearly all their wages; 3) a world in which mothers and fathers could choose to work part time until their children are in primary school without changing employers or losing health benefits; 4) a world in which the normal workweek was 35–37 hours, and parents could take occasional days off with pay to attend to family needs; 5) a world in which all parents have the right to place their children in high-quality day care provided by well-educated professionals; 6) a world in which this child care is provided at no cost or very low cost to parents.

Such a world is a reality in several European countries, where public policies distribute the cost of caring for children across society. Such an investment in the United States would be equivalent to 1.7 percent of GDP (about half of what we now commit to public education), and is justified because children are "public goods."

WORK/COUPLES/CAREERS (October 2003)

It's About Time: Couples and Careers. Edited by Phyllis Moen (Professor of Sociology and Life Course Studies, Cornell University). Ithaca, N.Y.: ILR Press (Cornell University Press), March 2003, 436 p.

Americans are all experiencing radical—and uneven—transformations in the nature of careers, work, families, communities, gender values, and the life course. Today, most workers are married to other workers, and most children have working mothers. This reconfiguration of the contemporary work force, along with increasing longevity, aging baby boomers, technological change, and globalization, is at odds with the script of traditional life courses of everyone.

This book presents the findings of the *Cornell Couples and Careers Study* (Moen, *et al.*; Cornell Employment and Family Careers Institute, 1999), on how members of two-earner families manage their paid work careers along with their conjoint family career. Chapters cover such topics as work-hour strategies (time is always an issue), competing clocks, timing parenthood, journey to work, managing households, turning points in work careers, factors that predict success, prioritizing careers, the new technology climate and the rise of tele-commuting, alternative employment arrangements, moving toward retirement (the interest of older workers in work-hour reductions), the case of same-sex couples, work-life integration, and family-friendly policies.

It's about time for the United States to confront the realities and needs of contemporary working couples, indeed all members of the new work force. To do so requires more than Band-Aid, short-term remedies. Some necessary policy ingredients: 1) employers should clarify, enforce, and further develop existing rules and regulations on time and career path flexibilities, and foster a culture supporting alternative work-hour and career options; 2) employers and governments should keep track of the variations in work-hour and career path arrangements among their own work forces and in the nation at large; 3) visionary employers should recognize the growth of the new and more heterogeneous work force, and change the outdated assumptions of linear, full-time, lockstep career paths; 4) risks and transitions should be seen as human resource, work force, and labor issues; 5) most important, the United States needs a comprehensive re-assessment of the hidden and outdated assumptions embedded in the Fair Labor Standards Act, Unemployment Insurance, Social Security, payroll taxes, and welfare—especially rules and regulations that constrain worker options.

[NOTE: Complements *Families That Work* (page 168).]

WORLD/HUMAN EVOLUTION (January 2004)

The Future Evolution of Man: The Divine Life upon Earth. Sri Aurobindo Ghose (1872–1950). Compiled with a summary by P.B. Saint-Hilaire. Twin Lakes, Wisc.: Lotus Press, September 2003, 148 p. (First published by Sri Aurobindo Ashram, Ondicherry, India, 1963.)

The nine chapters concern: 1) The Human Aspiration: man's highest aspiration would indicate the gradual unveiling of the Spirit within, the preparation of a higher life upon earth; 2) The Place of Man in Evolution: an evolution of consciousness is the central motive of terrestrial existence; "man occupies the crest of the evolutionary wave; with him occurs the passage from an unconscious to a conscious evolution ... a change of consciousness is the major fact of the next evolutionary transformation"; 3) The Present Evolutionary Crisis: reason cannot arrive at any final truth because it can neither get to the root of things nor embrace their totality; we must seek spirituality, restored to its true sense, for the directing light and the harmonizing law; the present crisis comes from "a disparity between the limited faculties of man—mental, ethical, and spiritual—and the technical and economic means at his disposal; 4) Standards of Conduct and Spiritual Freedom: four main principles successively govern human conduct: personal need, the good of the collectivity, the ethical ideal, and the expression of the Divine within us; 5) The Development of the Spiritual Man: spirituality is a progressive awakening to the inner reality of our being; 6) The Triple Transformation: a shifting of consciousness, opening to an Infinity above us, the supramental transformation; 7) The Ascent Toward Supermind: the transition from mind to Supermind is a passage from Nature into Supernature; the four steps of the ascent are Higher Mind, Illumined Mind, Intuitive Mind, and Overmind; 8) The Gnostic Being: the supramental or Gnostic being will be the perfect consummation of the spiritual man; the Gnostic consciousness will process toward an integral knowledge; 9) The Divine Life upon Earth: the law of the divine life is universality in action, with the sense of the true oneness of all; new powers of consciousness will develop in the Gnostic being who will use them both for knowledge and for action.

[NOTE: Sri Aurobindo Ghose is the author of many works on evolution and yoga, reprinted in many editions. His is perhaps the first use of the phrase "conscious evolution" used by some futurists. An interesting spiritual contrast to the technologically enhanced humans considered below.]

SOCIETY/LEADERSHIP (November 2003)

Our Culture of Pandering. Paul Simon (Director, Public Policy Institute, Southern Illinois University). Carbondale, Ill.: Southern Illinois University Press, October 2003, 208 p.

Former U.S. Senator (D-IL) Simon asserts that almost all big strides forward came because a few leaders defied public opinion. But we have spawned "leadership" that does not lead, that panders to our whims rather than telling the truth, that follows the crowd rather than challenges us. "*Pandering is not illegal, but it is immoral. It is doing the convenient when the right course demands inconvenience and courage.*" Leaders in all areas are guilty of giving in to what is easy instead of fighting for what is right.

1) Politics: to the extent that candidates follow the polls, they will sound alike; attention is given to civil liberties ("freedoms are easier to give away than to protect"), Social Security (we need leaders who aren't afraid of taking on monsters), taxes (the greatest form of government waste is payment of interest on the national debt, which could be avoided if politicians stop pandering with tax cuts), crime (few candidates suggest we should be smart on crime as well as tough; the United States has twenty-five percent of the world's prisoners and would be better off if nonviolent criminals spent thirty to sixty days in prison and were then assigned to serve society in a homeless shelter or some other assignment that will do them more good and cost less money), gambling (the only addiction promoted by government, Americans spend six percent of their national income on gambling vs. eight percent on groceries);

2) The Media: which offers the pious but lame defense "we're giving the public what it wants"; attention is given to too much violence and what the media is not covering well (poverty, religion, world affairs, presidential campaigns);

3) Religion: the great sin of faith leaders is the tendency to comfort and pander, but not to build bridges to other faiths and peoples or help the most miserable; a willingness is needed to stand up and do the extra "little things";

4) Education: much-needed changes include a longer school year (U.S. schools average 180 days/year, in contrast to 280 days in Singapore, 243 in Japan, and 240 in Germany), universal preschool education (only thirty percent of U.S. three-year-olds received schooling or custodial care, in contrast to one hundred percent in France and ninety-five percent in Belgium), promotion of adult literacy (23 million U.S. adults cannot read a newspaper), more emphasis on foreign languages and study abroad, and addressing the "excessive use and abuse of academic jargon" in higher education that meets publish or perish criteria but contributes nothing (this pandering takes academicians and their students away from the real world).

> [NOTE: Fresh, thoughtful, and sensibly inspiring in the public interest, much like the late John W. Gardner. Although pandering by politicians and the media is obvious, Simon bravely takes on the much less obvious realms of education at all levels and religion.]

SOCIETY/COMMUNITY (November 2003)

Better Together: Restoring the American Community. Robert D. Putnam (Professsor of Public Policy, Harvard University; founder, the Saguaro Seminar) and Lewis M. Feldstein (President, New Hampshire Charitable Foundation). New York: Simon & Schuster, September 2003, 318 p.

Inspiring case studies of efforts resulting in building social capital:

* Valley Interfaith in the Texas Rio Grande Valley (a decades-long community organizing effort, bringing basic services to the mostly Spanish-speaking residents);

* Branch Libraries in Chicago (the new library is an active and responsive part of the community to help link and change neighborhoods);
* The Shipyard Project in Portsmouth, New Hampshire (a participatory dance project to explain the shipyard to the city);
* The Dudley Street Neighborhood Initiative in Boston (a civic association that overcame ethnic differences);
* The Tupelo Model (after half a century of hard work, Lee County, Mississippi, has been dramatically transformed);
* Saddleback Church in Lake Forest, California (a highly successful evangelical Christian mega-church with some forty-five thousand members and a seventy-four-acre complex of facilities);
* Do Something (a national organization to encourage community activism and develop leadership skills in young people);
* Harvard Union of Clerical and Technical Workers (a new kind of union with the explicit goal of deepening relationships between labor and management);
* The Experience Corps in North Philadelphia (volunteers who help to raise the ambitions of poor kids);
* United Parcel Service (third in the list of Top Ten companies for diversity in 2001, UPS builds its operation on cooperation, connections, and loyalty within its 370,000 member work-force);
* Craigslist in San Francisco (a successful on-line community connected to the SF Bay Area);
* And a "Positive Epidemic of Civic Engagement" in Portland, Oregon (over the past three decades, Portland has one of the highest levels of civic participation in the United States).

Virtually none of these efforts set out "to build social capital," although *"connective strategies" are the shared feature in all of these cases.* What distinguishes these efforts from others is that the protagonists understand and emphasize the centrality of relationships and interpersonal connections. The benefits of social capital spill beyond those immediately involved and can be used for many other purposes (for example, the more neighbors who know one another by name, the fewer crimes a neighborhood will suffer). Misguided public policies can weaken or destroy social capital, but public power is often a necessary ingredient in building community.

So "*the argument sometimes heard that civil society alone can solve public issues if only the state would get out of the way is simply silly. In the examples above, government support of participatory strategies was crucial.*" Also, smaller is better for forging and sustaining connections. Conversely, bigger is better for critical mass, power, and diversity that bridges social capital.

One way to combine the advantages of small scale with the offsetting advantages of large scope is federation, or nesting small groups within larger groups. "Bridging is not about 'Kumbaya' cuddling; it is about coming together to argue, as much as to share." Building social capital is neither all-or-nothing nor once-and-for-all. It is incremental and cumulative. "*Again and again, we find that one key to creating social capital is to build in redundancy of contact.*"

> [NOTE: Putnam is the author of *Bowling Alone: The Collapse and Revival of American Community* (S&S, 2000; *FS* 22:11/505), one of the most-cited social science books of the past decade.]

SOCIETY/NATIONAL SERVICE (November 2003)

United We Serve: National Service and the Future of Citizenship. Edited by E.J. Dionne Jr., Kayla Meltzer Drogosz, and Robert E. Litan (all Brookings Institution). Washington, D.C.: Brookings Institution Press, April 2003, 327 p.

Since September 11, respect for service soared as the United States forged a stronger sense of solidarity in the face of deadly enemies, and police and firefighters became the new heroes. The politics of national service were also transformed. "*In the first nine months after September 11, applications for America-Corps jumped 50 percent, those for the Peace Corps doubled, and those for Teach for America tripled.*"

Even before the attacks of September 11, President Bush had signaled a warmer view of service than many in his party. After September 11, service became a stronger theme in the president's rhetoric. In his 2002 State of the Union message, he called on Americans to give two years of service to the nation over their lifetimes and announced creation of the USA Freedom Corps (a post-9/11 gloss on older ideas).

There is also a new acknowledgement across the political divide that government support for volunteers can provide essential help for valuable institutions that we too often take for granted. That national service has become a bipartisan goal is an important achievement. "*The debate over national service is a debate over how we Americans think of ourselves. It is a debate over how we will solve public problems and what we owe to our country and to each other.*"

The thirty-eight brief essays herein, most of them original or substantial adaptations, include:

1) How September 11 Changed Us: Robert D. Putnam on restored civic engagement ("Bowling Together"), Theda Skocpol on whether the reinvigorated sense of the American "we" may gradually dissipate, Charles Moskos on recruitment for America's all-volunteer force as a continuing problem despite all the "patriotism-lite" flag waving ("*if we want to encourage the ideal of service, there must be a push to link federal college aid to enlistment, whether military or civilian*");

2) Politics of the Service Debate: Harris Wofford on AmeriCorps, the Bush Administration's civic agenda, Sen. John McCain on patriotism as reaching beyond our self-interest, Bill Clinton on the duties of democracy, four ways to scale up national service (recruit more citizen-soldiers, expand AmeriCorps and Senior Corps, link national service and homeland security, replace the Federal Work-Study Program with Serve-Study), establishing a National Citizenship Day to promote service;

3) Universal Service? Robert Litan on the case for universal service for all young people, Bruce Chapman countering that this is "a bad idea whose time has passed" (a case study in unintended consequences; an invitation to scandal; prohibitive costs), Litan's reply to Chapman (there are ample opportunities for young people to make important contributions to unmet social needs, with many personal and social benefits), a proposal by Michael Lind for a new kind of national service based on a citizen militia at the local level, Rep. Charles B. Rangel on reinstating the draft and a critical reply by Caspar W. Weinberger.

Other essays discuss experiences of service, the challenges of civic participation, religion and civil society, and how people learn to be civic.

> [NOTE: Excellent collection of essays suggesting the many debates and possibilities regarding this durable and evolving idea. The problem is settling on which course to pursue.]

WORLD/HUMAN RIGHTS (January 2004)

Ethnic Conflict in World Politics. 2nd Ed. Barbara Harff (Professor of Political Science, U.S. Naval Academy) and Ted Robert Gurr (Distinguished Professor, University of Maryland). Boulder, Colo.: Westview Press, January 2004, 237 p.

Since the end of World War II, protracted conflicts over the rights and demands of ethnic and religious groups have caused more misery and loss of human life than any other type of local, regional, or international conflict. These conflicts are also the source of about two-thirds of the world's fifteen million refugees in 2002 ("at least twice as many others have been internally displaced by force and famine"). Violent conflicts between communal groups and states increased from the 1960s to the early 1990s, but turned sharply downward thereafter.

At least as many communal groups are still making ethnopolitical claims at the beginning of the twenty-first century as in the early 1990s. But their leaders are more likely to pursue group interests by conventional political means, often with international support, and are scoring significant gains (e.g., South Africa's largely peaceful transformation to a multiracial democracy, and Germany's acceptance of Turks and other immigrants as citizens).

Still, "assertion of group identities continues to be a major factor in the politics of almost every world region," and demands that group rights be recognized, self-government granted, and public resources spent on disadvantaged groups "will be major forces in domestic and international politics for decades to come."

Two other factors are likely to reinforce future ethnic challenges: growing inequalities within and between societies, and the rising assertion of religious identities (for example, Hindu militants attacking Muslims in India).

Five principles for managing ethno-political conflict: 1) states and civil society should recognize and promote the rights of minorities; 2) democratic institutions and power sharing are the best means for protecting group rights; 3) conflicts over self-determination are best settled by negotiations for autonomy within existing states; 4) international actors should protect minority rights and promote settlement of ethno-political wars; 5) international actors may use coercive means to stop civil wars and mass killing of civilians.

Annotated Bibliography (J–N)*

Only a few of the many books dealing with the future can be listed here. Selection criteria include recency of publication, trustworthiness of the author, readability, and usefulness to nonspecialists. Certain classic works have also been included, as well as a few other books not fully meeting the selection criteria but still likely to be of interest to readers. Information on the most recent books is available on the World Future Society's Web site (wfs.org), which also provides information through Future Survey*—the best available guide to current literature dealing with the future.*

* Part 3 of four parts, one in each volume in this series. Reprinted with permission from Edward Cornish and the World Future Society.

Jensen, Rolf. *The Dream Society: How the Coming Shift from Information to Imagination Will Transform Your Business.* New York: McGraw-Hill, 1999.

The director of the Copenhagen Institute for Futures Studies argues that the future of business lies not in selling products but in selling dreams and emotions. He identifies six markets: 1) the market for adventures; 2) the market for love, friendship, and togetherness; 3) the market for care (for example, pets); 4) the who-am-I market (products that proclaim their owner's identity, such as fashion, automobiles, and accessories); 5) the market for peace of mind; 6) the conviction market, which is also referred to as cause-related marketing ("green" products, worker welfare, etc.).

Johnston, William B., and Arnold H. Packer. *Workforce 2000: Work and Workers for the 21st Century.* Indianapolis, Ind.: Hudson Institute, 1987.

This study, prepared for the U.S. Department of Labor, examines the forces shaping the American economy, such as the integration of the global economy, the shift from goods to services, and the proliferation of advanced technologies. Challenges for policy makers include finding ways to accelerate productivity increases in service industries, maintain the dynamism of an aging workforce, and reconcile the conflicting needs of women, work, and families.

Jones, Glenn R. *Free Market Fusion: How Entrepreneurs and Nonprofits Create 21st Century Success*. Denver, Colo.: Cyber Publishing Group, 1999.

Educational entrepreneur Jones, creator of Knowledge TV and the University of the Web, argues that the recipe for a strong economy in the twenty-first century will be to mix the unique capabilities of for-profit and nonprofit organizations. The book includes interviews with futurists Theodore Modis and Alvin and Heidi Toffler.

Jungk, Robert, and Johan Galtung, eds. *Mankind 2000*. London: Allen and Unwin, 1969.

This volume contains thirty-five papers from a 1967 conference on the future in Oslo. It is now mainly of historic interest for what it shows about futurist thinking when the movement was getting started.

Jungk, Robert, and Norbert Müllert. *Future Workshops: How to Create Desirable Futures*. London: Institute for Social Inventions, 1987.

Jungk ran his first "future-creating workshop" in 1962. Ideally, he decided a workshop lasts three days. Day one is devoted to a critique of the situation to be addressed. Day two focuses on fantasizing about ways to solve the problems being addressed. Day three reviews suggested solutions for practicality and getting action started. The editors conclude that the workshops are a remarkable tool for harnessing the creative forces within society.

Kahn, Herman, and Anthony J. Wiener. *The Year 2000: A Framework for Speculation on the Next Thirty-Three Years*. New York: Macmillan, 1967.

This volume summarizes the thinking of Herman Kahn and his Hudson Institute colleagues concerning the world of the future. At the time of its appearance, many futurists regarded the book as possibly the most impressive work then available in terms of its disciplined and penetrating attempt to identify and describe the major trends in Western society. The book describes the "basic, long-term multifold trend" in Western civilization, and projects economic and other trends to the end of the twentieth century.

Kahn, Herman, William Brown, and Leon Martel. *The Next 200 Years: A Scenario for America and the World*. New York: William Morrow and Company, 1976.

A highly optimistic view of America's future: America will become increasingly wealthy, and the problems associated with shrinking supplies of fossil fuels and increasing pollution can be overcome.

Kaku, Michio. *Visions: How Science Will Revolutionize the 21st Century and Beyond*. New York: Doubleday, 1997.

Physicist/science writer Kaku bases his book on interviews with more than 150 top scientists in many fields. He expects major research projects already under way to yield dramatic results between now and the year 2020, and these new findings should make possible wholly new technologies in the decades to 2050 and beyond.

Kelly, Kevin, Peter Leyden, and Members of the Global Business Network. *What's Next?* Cambridge, Mass.: Perseus Publishing, 2002.

Selected quotes from twenty-two visionary scientists, business leaders, and futurists at a computer conference are organized into topic areas, including geopolitics, values and belief systems, science, technology, environment, civilization, and more.

Kiernan, Matthew J. *The Eleven Commandments of 21st Century Management*. Englewood Cliffs, N.J.: Prentice Hall, 1996.

Tomorrow belongs to the smaller, more agile companies.

Managers who are not already working in one had better start now to position their companies to succeed in a new business environment demanding constant innovation and creativity. Among the new commandments for managers: get innovative or get dead; use all of your people, all of their skills, all of the time; and turn organizational learning into a corporate religion.

Kindleberger, Charles P. *Manias, Panics, and Crashes: A History of Financial Crises*. Rev. ed. New York: Basic Books, 1989.

This book is an excellent investment in your financial future. Read it before you start investing.

Kressley, Konrad. *Living in the Third Millennium*. Mobile, Ala.: Factor Press, 1998.

A professor of political science offers a readable, straightforward vision of probable futures, with an emphasis on helping readers develop skills for managing their own futures. Topics include the art of forecasting, planning your career, preparing for financial security, becoming proactive about your health, and more.

Kuhn, Thomas. *The Structure of Scientific Revolutions*. Chicago, Ill.: University of Chicago Press, 1970.

This book popularized the concept that science progresses by means of repeated "paradigm shifts," such as when the Copernican view of the solar system replaced the Ptolemaic view. Since then, the term "paradigm shift" has become a catch term in business, the social sciences, and elsewhere.

Kurian, George, and Graham T.T. Molitor. *The Encyclopedia of the Future*. 2 vols. New York: Macmillan, 1996.

Graham Molitor, vice president of the World Future Society, spent five years recruiting distinguished futurists and other scholars to produce this monumental work. Probably never before had so many scholars collaborated on a future-oriented project. The result was an enormous assemblage of ideas and insights about the future. The four hundred separate articles discuss the future of everything from comic books to capitalism, from dentistry to demography, peacekeeping and the performing arts. Though now dated, it contains an enormous amount of material that is still useful and interesting.

———. *The 21st Century. Macmillan Compendium.* New York: Macmillan, 1999.

This one-volume edition of the *Encyclopedia of the Future* condenses the contents of the two-volume set.

Kurzweil, Ray. *The Age of Spiritual Machines: When Computers Exceed Human Intelligence.* New York: Viking, 1999.

A noted inventor describes likely advances that will result in computers exceeding the memory capacity and computational ability of the human brain by the year 2020; relationships with automated personalities who will be our teachers and companions; and information fed directly into human brains along direct neural pathways. Eventually, the differences between humans and computers will be so blurred that we will believe the machines are conscious when they say they are.

Lee, Laura. *Bad Predictions.* Rochester, Mich.: Elsewhere Press, 2000.

A compendium of erroneous forecasts from many sources. Covers transportation, technology, medicine, arts, business, history, and more. Demonstrates that anyone can make a mistake when trying to predict, and many forecasts are highly amusing in hindsight. Includes introduction but no commentary on individual forecasts.

Linstone, Harold A. *Decision Making for Technology Executives: Using Multiple Perspectives to Improve Performance.* Norwood, Mass.: Artech House, 2000.

The author, a veteran of many years in technical planning for the Hughes and Lockheed corporations, astutely analyzes the problems of making decisions concerning technology:

Specific examples discussed include such disasters as the Maginot Line, the Exxon *Valdez*, and Three Mile Island. Though somewhat technical, this well-informed text requires no deep knowledge of technology or mathematics to appreciate its important insights into decision making.

Lippitt, Lawrence L. *Preferred Futuring: Envision the Future You Want and Unleash the Energy to Get There*. San Francisco, Calif.: Berrett-Kohler, 1998.

The Preferred Futuring concept, developed by Ron Lippitt (the author's father) and Ed Lindaman in 1968, focuses on getting all the stakeholders together to develop a vision that is clear, detailed, and generally understood. The vision then is translated into action goals, a series of planned steps with accountability identified, together with a structure to implement the action plan.

[Note: This approach resembles the Future Search approach described by Marvin R. Weisbord and Sandra Janoff. See entry for their book in volume four.]

McGuire, Bill. *A Guide to the End of the World: Everything You Never Wanted to Know*. New York: Oxford, 2002.

An authoritative yet highly readable book that will open readers' eyes to the dangers posed by our natural environment including climate change (global warming versus a new ice age), super-volcanic eruptions, giant tsunamis, earthquakes, asteroids, and comets. The author, a professor of geophysical hazards at University College London, provides much eye-opening new information that scientists have developed. An exceptional book on this topic.

McNeill, William H. *Plagues and Peoples*, 1976. Rev. ed. New York: Anchor Books/Doubleday, 1998.

AIDS, SARS, and Mad Cow Disease demonstrate the continuing importance of epidemics in shaping events. McNeill's classic text traces the impacts of epidemics on human life through the centuries.

Maddox, John. *What Remains to Be Discovered: Mapping the Secrets of the Universe, the Origins of Life and the Future of the Human Race*. New York: Free Press, 1998.

A former editor of the British science journal *Nature* takes a serious look at what scientists don't know. "The record of previous centuries suggests that the excitement in the years ahead will spring from the answers to the questions we do not yet know enough to ask ... The problems that remain unsolved are gargantuan. They will occupy our children and their children and so on and on for centuries to come, perhaps even for the rest of time."

Marien, Michael, and Lane Jennings, eds. *What I Have Learned: Thinking About the Future Then and Now*. Westport, Conn.: Greenwood Press, 1987.

Leading futurist authors of the 1970s and 1980s report how their thinking about the future has changed. Contributors include W. Warren Wagar, Kenneth E. Boulding, Willis W. Harman, Victor Ferkiss, Irene Taviss Thomson, Robert T. Francoeur, Jim Dator, Amitai Etzioni, Walter A. Hahn, Joseph F. Coates, Vary T. Coates, Harold A. Linstone, Bertram Gross, Kusum Singh, and Hazel Henderson. The varied contributions demonstrate the wide diversity of views among people sharing a common interest in the human future. This diversity may be frustrating to people expecting to be told exactly what the future will be like by a knowledgeable expert, but the diversity of the contributions also shows the creative strength of the futurist community, which provides an opportunity for futurists with strongly opposed ideas to come together and share their often radically different ideas in a friendly and mutually respectful venue.

Marsh, Nick, Mike McAllum, and Dominique Purcell. *Strategic Foresight: The Power of Standing in the Future*. Melbourne, Australia: Crown Content, 2002.

An Australian perspective on introducing strategic foresight into organizations. The book also discusses national foresight programs in New Zealand, Finland, Brazil, and elsewhere.

Masini, Eleanora. *Why Future Studies*? London: Grey Seal Books, 1993. (Out of print. Copies still available through the World Future Society.)

An Italian futurist and professor examines the history, principles, concepts, philosophy, and ethical elements of the futures field. Masini spent many years working for the World Futures Studies Federation and teaching at Rome's Pontifical Gregorian University. Her long experience is reflected in this thoughtful book.

Mason, Colin. *The 2030 Spike: Countdown to Global Catastrophe*. London and Springfield, Va.: Earthscan Publications, 2003.

This is a recent example of a "doomsday" book anticipating global calamity unless action is taken immediately. The author, a well-known Australian journalist and politician, believes that depleted fuel supplies, massive population growth, poverty, global climate change, famine, growing water shortages, and international lawlessness will converge in the 2030 decade and may smother civilization.

Masuda, Yoneji. *The Information Society as Post-Industrial Society*. Bethesda, Md.: World Future Society, 1981.

A principal architect of Japan's $65 billion computer-usage plan offers a vision of the society of the future when computers will free people to live more creative and happy lives. He discusses computer-controlled vehicle systems, automated supermarkets, etc. Though it is now dated, Masuda's visionary book excited many in the computer world when it first appeared, and his ideas are still interesting.

Matathia, Ira, and Marian Salzman. *Next: Trends for the Near Future*. Woodstock, N.Y.: Overlook Press, 1999.

Trend watchers for a New York advertising firm offer a host

of predictions for business, technology, and lifestyle changes in the years just ahead. The book is packed with facts and forecasts but lacks references to the original materials. The text seems largely aimed at marketers, and the authors are not shy about advertising their research services.

May, Graham H. *The Future Is Ours: Foreseeing, Managing and Creating the Future*. Westport, Conn.: Praeger, 1996.

A lecturer at Leeds Metropolitan University in England explores such questions as why we forecast the future despite the likelihood we will be wrong, how people relate to the future, and other issues. This is a college-level text for use in future-oriented courses in business, management, urban planning, and other areas.

Mazaar, Michael. *Global Trends 2005: An Owner's Manual for the Next Decade*. New York: St. Martin's Press, 1999.

Mazaar, a fellow of the Center for Strategic and International Studies in Washington, D.C., offers a general overview of the transformations to be expected in the near future. He also considers a few surprise scenarios, such as a media-savvy demagogue who rises to power offering alienated millions an authority figure to replace traditional institutions that have been undermined by information overload.

Meadows, Donella H., Dennis L. Meadows, Jorgen Randers, and William W. Behrens III. *The Limits to Growth*. New York: Universe Books, 1972.

This volume summarizes the Club of Rome report prepared at MIT by Dennis Meadows and his colleagues, using the computerized system dynamics approach developed by Professor Jay Forrester. The pessimistic report created a sensation in intellectual circles, especially in Europe, with its contention that, if present patterns of rapid population and capital growth are allowed to continue, the world faces "a disastrous collapse."

Merriam, John E., and Joel Makower. *Trend Watching: How the Media Create Trends and How to Be the First to Uncover Them.* New York: AMACOM, 1988.

This how-to book deals with understanding, analyzing, and anticipating trends based on what is being reported in the media.

Michalko, Michael. *Cracking Creativity: The Secrets of Creative Geniuses.* Berkeley, Calif.: Ten Speed Press, 1998.

A clear and lively discussion of creativity techniques.

Miller, James G. *Living Systems.* New York: McGraw-Hill, 1978.

A mammoth volume presenting an overview of the author's theory of living systems at both high and low levels of complexity. He creates a hierarchy of seventeen components of systems.

Millett, Stephen M., and Edward J. Honton. *A Manager's Guide to Technology Forecasting and Strategy Analysis Methods.* Columbus, Ohio: Battelle Press, 1991.

A short, well-informed, and practical guide to the use of forecasting and strategy analysis methods in corporate planning. It provides forthright and detailed critiques of their strengths and weaknesses.

Mintzberg, Henry, Bruce Ahlstrand, and Joseph Lampel. *Strategy Safari: A Guided Tour Through the Wilds of Strategic Management.* New York: Free Press, 1998.

A primer on business strategy with a critique of different approaches.

Modis, Theodore. *An S-Shaped Trail to Wall Street: Survival of the Fittest Reigns at the Stock Market.* Geneva: Growth Dynamics, 1999.

In his earlier books *Predictions* (Simon & Schuster, 1992) and *Conquering Uncertainty* (McGraw-Hill, 1998), physicist/

futurist Modis showed how scientific theories about seasonal change and cycles of behavior in plants and animals might be used to explain—and forecast—changes in human society. This newer book uses this approach to explain events in the stock market. The book offers fascinating insights into parallels and between completely unrelated areas of nature and human society.

Molitor, Graham T.T. *The Power to Change the World: The Art of Forecasting*. Potomac, Md.: Graham T.T. Molitor, 2003.

A spiral-bound presentation of the "Molitor Model of Change: Basic 22 Signature Patterns of Change." The book features more than two hundred charts showing trends of many varieties. The patterns of change include litigation, random phenomena, subtle impacts, catalysts, voluntary accommodation, and much more. An unusual and stimulating presentation of material. Provenance of the original data is not indicated.

Moravec, Hans. *Robot: Mere Machine to Transcendent Mind.* Oxford, U.K.: Oxford University Press, 1998.

A former Carnegie Mellon University robotics expert believes robots will model themselves after successful biological forms. He anticipates that robots will become "intelligent machines" that will learn our skills, share our goals and values, and become our evolutionary heirs.

Morrison, Ian. *The Second Curve: Managing the Velocity of Change*. New York: Ballantine, 1996.

Help for managers preparing for future growth and change. "The second curve" is a revolutionary business model that allows companies to anticipate the rate of change, identify new directions, and know when to jump onto the second curve of change.

Mulhall, Douglas. *Our Molecular Future: How Nanotechnology, Robotics, Genetics, and Artificial Intelligence Will Transform Our World.* Amherst, N.Y.: Prometheus Books, 2002.

A readable, imaginative survey of cutting-edge technologies and their implications for human life. Discusses such things as transhumans, the possibility of a "singularity," molecular weapons, and scenarios for what may happen in the future.

Naisbitt, John. *Global Paradox: The Bigger the World Economy, the More Powerful Its Smallest Players*. New York: William Morrow, 1994.

The author argues that both nations and individuals are now breaking up into smaller and smaller units, because the telecommunications revolution is simultaneously creating a global economy and empowering its constituent parts, with power flowing particularly to small units. Goods and services can be sold all over the world with increasing ease, so the small nations and businesses are finding that they can compete more successfully than in the past.

———. *Megatrends: Ten New Directions Transforming Our Lives*. New York: Warner, 1982.

Naisbitt describes ten major trends affecting our society today. For instance, we are moving from an industrial society to an information society, from a national economy to a world economy, and from institutional help to self-help. Though now dated, this book was a best seller that got many people thinking about trends.

Naisbitt, John, and Patricia Aburdene. *Megatrends 2000*. New York: William Morrow, 1990.

Modeled on the best-selling *Megatrends*, this book focuses on a new set of perceived macrotrends, including the rise of the Pacific Rim countries, a religious revival, etc. This volume like its predecessor can be described as a "newsbook"—a book-length text designed to help time-short people understand the forces that give rise to news events. Despite critics sneering at this "millennial megababble," thousands of readers like Naisbitt's information-rich, readable presentation and have put his books on best-seller lists.

Naisbitt, John, with Nana Naisbitt and Douglas Philips. *High Tech—High Touch: Technology and Our Search for Meaning*. New York: Broadway Books, 1999.

This highly readable and informative but loosely organized book argues that people need "high touch" experiences—the non-technological joys of living—to set off the high-tech aspects of their lives. The book includes a strong chapter on the social and ethical consequences of genetic technology.

Nanus, Burt. *Visionary Leadership: Creating a Compelling Sense of Direction for Your Organization*. San Francisco, Calif.: Jossey-Bass, 1992.

This timely book explains what visionary leadership is all about and why it is important to develop the skills necessary for leading organizations into the future. Leadership expert Burt Nanus, co-author of this best-selling book, shows you how to develop a vision, implement it, and know when it's time to "re-vision." A participant's workbook is also available.

———. *The Vision Retreat: A Facilitator's Guide*. San Francisco, Calif.: Jossey-Bass, 1995.

A leadership expert offers a logical, step-by-step process for creating and implementing a new direction for your organization.

Nirenberg, John. *Power Tools: A Leader's Guide to the Latest Management Thinking*. Englewood Cliffs, N.J.: Prentice Hall, 1997.

A guide to one hundred of the latest management "tools," with an assessment of just how useful they are.

Notes on Contributors

Arnold Brown is chairman of Weiner, Edrich, Brown, Inc., one of the oldest and most prominent trend analysis consulting firms in the world. His company's clients have included many of the largest corporations, as well as governments and non-profits. He is a coauthor with his partner of three well-known books on change and how to benefit from it: *Supermanaging*; *Office Biology*; and *Insider's Guide to the Future*.

Linda Brown, formerly a director of senior social services, science educator, workforce developer, and academic adviser, has a special interest in improving the world for all generations. Since earning a master's degree in Studies of the Future at the University of Houston-Clear Lake, she has been reporting to corporate clients on the outlook for American families.

Marcus Bussey is a teacher and futurist. He has worked with community groups and a variety of small schools in order to generate positive visions of schooling and social reclamation of the educational process. He is also a musician and writer who has published more than thirty journal articles and book chapters on education, culture, and creative engagement with life. He can be reached at fitzbus@ozemail.com.au.

Josh Calder is a professional futurist in Washington, D.C., at the consulting firm Social Technologies. He studies societal and technological change and helps corporations and governments prepare for and shape the future. He is the creator of Projections: A Futurist at the Movies, (www.futuristmovies.com), which examines the futures depicted in film. Josh can be reached at jcalder@erols.com.

Marvin Cetron, Ph.D., has been one of the world's most respected forecasters for more than forty years, having developed many of the techniques now used throughout the profession. He has consulted for more than 400 of the *Fortune* 500 corporations, 150 professional and academic organizations, and 100 departments within the U.S. and foreign governments. He also has served as an advisor to the White House in every administration from John F. Kennedy to Bill Clinton. He has long-standing relationships with several major clients in the travel and hospitality industry, and currently publishes a regular column for *HSMAI Review* (Hospitality Sales & Marketing Association International). He has published thirty-eight books.

Joseph F. Coates is a consulting futurist and coauthor of *2025: Scenarios of U.S. and Global Society Reshaped by Science and Technology*. His address is 3738 Kanawha Street, NW, Washington, DC 20015. Telephone: 1-202-363-7440; e-mail: joe@josephcoates.com, and Web site: www.josephcoates.com.

Ann Coombs is a Thought Leader, Futurist, and President of Coombs Consulting, Ltd. She speaks internationally on innovative thinking, creativity, and being a "rule breaker" in the corporate world. Ann is the author of the best-seller *The Living Workplace—Soul, Spirit and Success in the 21st Century* (Harper Collins, 2001). Her newsletter on *living* workplaces can be found at www.thelivingworkplace.com, which provides current trends and research on the global workplace, and is read by 250,000 readers worldwide.

Marilyn Dudley-Rowley, Ph.D., teaches sociology, political science, and criminal justice courses at Sonoma State University near San Francisco, California. A wide sampling of prior occupations gives her multiple perspectives on the human condition. After a tour-of-duty in the U.S. Army, she hosted and produced radio news shows, did investigations for a number of federal task forces, performed research that helped tailor public lands legislation, worked as an archaeologist and geoscientist, drove as a long-haul trucker cross-country, worked as a clinical counselor and marriage and family therapist, and participated in a space station simulation in Russia. She is affiliated with NASA, the Department of Defense's Human Factors Engineering Technical Advisory Group, the World Affairs Council of Northern California, the American Institute of Aeronautics and Astronautics, and other organizations. E-mail: md-r@ops-alaska.com; Web site: www.ops-alaska.com.

Richard Gerson, Ph.D., CPT, CMC, is a sports and performance psychologist and is president of The HEADcoaching Institute, which provides programs that train your brain for maximum gain in sports, business, and life. He is the author of nineteen books and more than four hundred articles, and is also a professional speaker. His e-mail is headcoach@headcoaching.com.

Nat Irvin II, D.M.A., President of Future Focus 2020 at Wake Forest University's Babcock Graduate School of Management, works with organizations, particularly with young people, to examine social, political, economic, technological, and environmental issues that are expected to have the greatest effects on urban communities by the year 2020. Dr. Irvin also is Assistant Dean for MBA Student Development at Babcock, and a partner in Irvin, Goforth and Irvin, a training and communications consulting firm. He is in his tenth year as columnist for the *Winston-Salem (N.C.) Journal*, distributed by Media General Newspapers. He has also written for a host of magazines and other publications, including *Emerge Magazine*, and provides commentary for NPR's Weekend Edition.

Graham T.T. Molitor is president of Public Policy Forecasting, and vice president & legal counsel of the World Future Society. He headed lobbying staffs at General Mills and Nabisco, chaired a legislative Commission on the Future, directed research for the White House Conference on the Industrial World Ahead, served on the White House Social Indicators Committee, headed research for both of Vice President Nelson Rockefeller's presidential campaigns as well as part-time roles in two other presidential campaigns, worked as legal counsel in the U.S. Congress, and served with the Assistant Chief of Staff at the Pentagon. Publications include: *The Power to Change the World: The Art of Forecasting*, 2003; *The 21st Century* (coeditor), 1999; and *The Encyclopedia of the Future* (coeditor and editorial board chairman), 1996.

Caleb Rosado, Ph.D., is director of the graduate program in Urban Economic Development at Eastern University, Campolo College of Graduate and Professional Studies, Philadelphia, Pennsylvania. He is a sociologist, an inspirational motivational speaker, educator, facilitator, and consultant. Dr. Rosado holds a doctorate in sociology from Northwestern University and has published three books and numerous professional articles. His current research is in three areas: the impact Latino congregations are having in their respective communities, why Latinos are not making it in school, and Spanish mystics. His latest publication is "Affirmative Action: A Time for Change?" in *Workforce Diversity, Vol. 1: Concepts and Cases* (ICAFAI University Press, Hyderabad, India). E-mail: calebrosado@earthlink.net; Web site: www.rosado.net.

Arthur B. Shostak, Ph.D. (Editor), holds the title of Emeritus Professor of Sociology after recently retiring from Drexel University (Philadelphia, Pennsylvania), where he was a professor since 1967. Since he began college teaching in 1961, he has specialized in trying to apply sociology to real-time problems ("challenges") and in shaping and communicating long-range forecasts. While at Drexel, he directed a two-year study of teenage attitudes toward the world of work and related matters. He has written, edited, and coedited more than thirty books and more than 160 articles, and was presented with the Pennsylvania Sociological Society's Distinguished Sociologist Award in 2004. He especially recommends his 2003 edited collection, *Viable Utopian Ideas: Shaping a Better World* (M.E. Sharpe, Armonk, N.Y.). You can contact him at shostaka@drexel.edu.

Index